GUIDE TO EMERGENCY SURVIVAL COMMUNICATIONS

---

by
Dave Ingram

Published and Distributed
by
Universal Electronics, Inc.
4555 Groves Road, Suite 12
Columbus, OH 43232, USA
(614) 866-4605
FAX (614) 866-1201

GUIDE TO EMERGENCY SURVIVAL COMMUNICATIONS

Author: Dave Ingram
Graphic Artist: Kenneth D. Prater

Publisher: Universal Electronics, Inc.
4555 Groves Road, Suite 12
Columbus, OH 43232
(614) 866-4605 Fax: (614) 866-1201

Direct all inquiries and orders to above address.

ISBN 0-916661-05-9
Printed in U.S.A.
Second Printing 1999

TABLE OF CONTENTS

PREFACE

1996 Power Outage Experience

As of late, there has been discussion of disaster communications, emergency power systems, and what to store for disasters. In the summer of 1996, a massive power outage hit many western states, including the one I'm in (I lost power), I kept careful logs of exactly what happened over the course of the initial few hours and into the next day of the outage. Here's the review.

The Event

A year ago, in the late afternoon an a fine summer day, a high-tension power line suffered a failure (reportedly due to a tree growing into the line) that resulted in an overload situation and then a cascading massive failure of the power grid in the Western U.S. The initial failure resulted in a widespread power loss to all areas. The failure subsequently knocked many substations off line and some utilities experienced generator shutdown (including nuclear reactors going off-line with multi-day restart delays). Due to damage and/or interlocks at substations, many smaller pockets were without power long after generating capacity was restored to the general system.

Chain Reaction

The power outage resulted in damage to both power-handling equipment and transmission lines, however, not always directly. A few substations were blown as the grid went down As many areas were left without power and lights were out, accidents involving vehicles skyrocketed. Some of those accidents involved collisions with power poles, eliminating the option of some areas having power even with a restored grid.

Observations

Although power was lost, I maintained a complete radio watch on TV, AM, FM, 2-meter Amateur and an assortment of public services frequencies, mostly on radios powered by AA batteries. I had an ample Supply of backup Lithium AA (10-year shelf life) and several sets of charged Ni-Cad rechargeable AAs. I had several solar AA chargers, so power was never a worry. I also had a couple 12-volt power packs with gel cells which run the amateur gear. After the event, I opted to pick up a larger 12-volt panel to charge the gel cells, too.

If you're worried about communications in a disaster, you need to worry about power. A long time ago, I settled on the AA battery rule—any equipment (flashlights, radios, Geiger counters, whatever) I purchased must be powered by AA batteries, except for high-power 12-volt radios, which use vehicle or 12-volt gel-cell power.

Keeping a stock of AA batteries in normal and rechargeable types—along with a small solar charger—eliminates any power issues for a long, long time (years).

Information Sources

First, monitoring TV stations was totally worthless—maybe even worse than worthless, as the information broadcast was simply wrong most of the time. Given a choice of repeating themselves or making it up, they'll make it up! All major network stations stayed on-line, but several smaller UHF stations went off the air at the moment of the outage, not to return. However, many lost their network feeds. One local television station activated their EBS system—you know, the 30-second tone followed by official instructions from the government (which I assume are along the lines of put jam in your pockets, you're about to be toasted).

AM radio, particularly talk radio stations (some of which were old civil defense stations proved to be the most current and up-to-date source of information on the outage. Probably because the producer's Rolodex is full of experts that have been on the radio before and because of the large number of dedicated listeners who call in reports for their neighborhood, just like real 2-way radio networks! I'd say that less than half of all local AM stations stayed on, but all of the talk radio/civil defense stations remained on the air. Specifically, callers reported stores that still had power, gas stations that were still pumping, fast food restaurants that were still serving. Later, emergency services provided a list of available air conditioned shelters. Note that all of this depended on the phone system working. Guests on these talk shows included power company spokes people and retired power company workers who explained the grid and likely causes of faults and problems that were occurring.

FM radio was OK, with some stations picking up feeds from AM stations, but unlike talk radio, most FM broadcasters were not prepared to handle anything. They did their best, which was loads better than TV but pathetic compared to AM radio. Ironically, far more FM stations, probably 80 percent, remained on the air. A few of the FM stations were acting in bizarre ways—the local NPR station had really weird pulses broadcast, probably due to a fault of remote site links or something of that nature.

All of the broadcast information was HIGHLY repetitious, with the same instructions for the sheepple—turn off your A/C, keep your refrigerator door closed, don't panic, DO NOT CALL 911. This cycled about every five minutes. It was worth checking every ½ hour to get any changes, but continuous monitoring was worthless.

Two-meter amateur radio was interesting. Although, several key local repeaters did have battery backup and stayed on, including a repeater network with phone patch, many repeaters lacked backup power and were down. Immediately after the outage, several self appointed repeater police hams—who had no connection with the repeater—were telling anyone checking in to get off the repeater to save power (which, in turn, used more power than the simple N7 message to which they were responding). In about half a hour, a local net was spontaneously created. Power workers who were hams had the correct information about the outage one hour before any other sources—including verification that the local nuclear power plant was safe and secure. Roaming hams provided a near-continuous survey of what was and wasn't working, somewhat like talk radio did, only with more detail and you could get specific questions answered.

Personal frustrations were that my radio was too underpowered (7 watts into a base antenna) to make it to the repeater that was selected for the net. If net control had polled for stations by area, rather than with a general call for check-ins, I could have been heard. As it was, I was walked over so I never managed to get my area into the net. Hams, pick up a power amplifier if you can; it'll help with simplex, too, If you're ever net control, please get check-ins by area! Because of this, I also added a proper antenna for my handheld radio.

A lot of "dummy" hams kept using the auto patch to call 911 to report the power outage (duh!). At least one repeater trustee actually turned off access to 911 as it was being abused (I'm sure a real emergency would have been put through or handled via a net.)

My scanner tuned to public service frequencies proved to be some help, however, there were many confused police and fire calls, accidents, etc., which yielded very little information about what was actually happening. The police were overwhelmed by accidents and burglar alarms which were triggered by the outage. Subsequently, they started ignoring the burglar alarms. I think a longer-term disaster would cause this source to be much more useful.

I did listen to shortwave and a few stations mentioned that something was happening on the West Coast, but that was it. Might be more useful for a broader, longer disaster. For CB to be useful in such a disaster, establish watch keeping with a friend via CB should this happen again. It would be advisable to count on CB to talk to people you know and have talked to before on CB with an appropriate plan for communicating.

Other Communications Systems

The telephone network remained largely operational, though a few areas were without service. Many offices that had AC-powered phone systems without backup were out, as were users of digital lines with no backup power.

The 911 system was overloaded within 10 minutes of the outage. Call rates jumped to 500 calls per hour for five hours (2500 calls) when normally 200 calls would come in—which is a 10 time increase. The dumber sheepple would call 911 for the two following reasons:

1. Question: My power is out, when will it be back on?
   Answer: Call the power company!

2. Question: The traffic light at xxx is out.
   Answer: Not an emergency!

The 911 operators specifically requested the media relay a message for people not to call 911 unless a life was in danger. One house burned down because the victims could not reach 911; fire fighters arrived after a passing police offer radioed the information that the house was on fire.

Two hours after the outage, cell phones and pagers continued to work, although getting a phone line was a challenge at times. Much to my dismay, the cell phone was more operational 2 hours after the event than the 2-meter autopatches I use, although several hours later that wasn't the case as some cell sites started to die out. About four hours into the outage, many backup systems started to drop off-line.

Consequences

Because of the outage, many things occurred, which resulted in the following conclusions:

1. There's a small number of really, really stupid people out there who screw things up and cannot survive even a couple of hours without help from someone.
2. There's a lot of dependency on some fairly weak service infrastructure—most people would be OK for a few days, but things could get bad very quickly in longer disaster situations.

Services Lost

1. Stores closed due to lack of power to run cash registers. Stores literally

wouldn't sell dry goods for cash because they were unable to scan the bar codes and run charge cards through their systems!

2. Gas Stations had no juice, no pumps. Only those with backup power were providing gas, most on a cash sale only—no cards.

3. ATM Machines - no cash because of power or phone problems.

4. Health Issues - lots of people depend on A/C and on electrical equipment like oxygen concentrators and needed to relocate to shelters when the power went out. Emergency services actually set up air-conditioned shelter lists and water/ ice supply points.

5. Water - some areas were without water because pumps had no power or pumping systems failed.

6. Sewage - some areas depend on sewage pumps to move the sewage uphill. You can guess what happened without power. Yuk!

Simultaneous Disasters

The power outage created an emergency services overload that resulted in minor incidents turning into major ones. About an hour after the outage occurred, a propane tanker overturned on a state highway in a small town that was a choke point between two regions in the state. The emergency services people, highway patrol and fire department, could barely deploy observers, let alone a force to solve the problem. As a result, all traffic between the two regions was halted except for a few people who knew how to get around back roads and shortcuts. Total nightmare that turned a one-hour delay into an eight-hour delay!

The news reported that the police went on tactical alert and that minor looting occurred in some small and remote communities in the western states.

Clueless People

As late as several hours into the outage, a lot of people, including the news people, dramatically misunderstood how widespread the power outage actually was. Specifically, their logic went along the lines of if the grocery store on this block has power, so the entire block must have power. Hence, this is a spotty, isolated outage. In fact, what happened is that everyone assumed that if a building had power, it was UTILITY power, which was simply not the case. Stores using backup generators caused mass confusion and underestimates of the severity of the power outage.

The Wave of Sheeple

Stores and gas stations with the foresight to have backup power became critical safety nets for what can only be termed hyper-short-term oriented people. Literally two hours after the outage, the talk radio stations were getting calls from hysterical people crying that they couldn't feed their children. Paraphrasing the problem statement from one called land I'm not kidding) "I have no food in my house. The fast food places are closed. The stores are closed. I have no money because the ATM machine is not working. I am almost out of gas because I had to drive around to find something that was open and now I cannot find a working gas station." Hungry kids were screaming in the background.

Two Words: Natural Selection

These people ended up getting help from the informed talk radio stations and ham operators who knew where things were operational. If it weren't for these resources, they would have either starved or called 911, screwing up 911 even more than they had already.

The Day After

Late into the second day, even though power was restored to some areas, a few areas that had experienced some sort of catastrophic failure (either a blown substation or power pole damage) were still without power. Ice and water distribution centers were set up and people turned into angry mobs. Classic quote from TV News: "Ice is like gold in the summer, ice and water can probably buy you some friends during a short-term disaster."

Lessons Learned

1. Being prepared and knowledgeable, you'll find that you will know more than the news media does in dealing with an emergency of this sort.
2. Having a lot of ice and water on hand is a good thing.
3. When something disastrous happens, stay at home until you know exactly what is happening and how to deal with the problem, then form your plans.
4. People who have experience doing something tend to do a lot better than those who don't.
5. Never trust anything they say on TV!

Things I Wished I Had During This Outage

1. A 12-volt solar panel with ample gel-cell batteries to power my radio equipment.
2. A 2-meter amateur transceiver with a suitable 2-meter power amplifier.
3. Proper antenna for my handheld radio and proper base station antenna for 2-meter band.
4. A 200- to 400-watt 115 VAC power inverter to be powered by a 12-volt deep-cycle auto battery. This inverter would run many pieces of equipment that I would have liked to have had working, such as the phone system, fax machine, small appliances and lights.
5. Proper 12-volt lights to be run off the deep-cycle battery.
6. A small generator with 800-watt capacity and ample fresh fuel.

by Anonymous
From the Internet

CHAPTER ONE

Introduction to Survival Communications

Survival is a basic instinct of every living creature. Universal, also, is the need to acquire accurate "what is happening" information and to communicate with others during emergencies and abnormal situations. A familiar example of those facts is preparing to "ride out" an evening thunderstorm by having candles, matches, and a battery-powered weather radio handy. An example of a more serious nature is installing a complete underground living shelter and communications center for surviving anything from a tornado to a nuclear disaster. Between those two extremes are numerous areas of survival preparedness applicable to our daily lives. Looking closely at these areas, four primary needs become apparent: food, water, protection, and communication with the "outside world." This book's focus is on the latter of these needs: communications. It is written in plain language using a minimum of technical jargon in hopes of helping everyone to be more "communications effective" during times of unusual events.

Survival communications can be applied to four open-ended categories: personal emergencies, weather disruptions, economic/political imbalance, and changes in world affairs. Each situation involves using various communications systems to inform or to be accurately informed of related events. Further, what works best for one situation (events occurring at a distance greater than 100 miles

Figure 1-1 *A wide array of communications systems and information resources address today's survival applications. Knowledge of various units capabilities and operation is the key to successful use. Shown here are: (A) a CB base unit, (B) a two-meter talkie, (C) GMRS handheld, (D) FRS handheld, (E) GPS handheld, (F) a shortwave receiver, (G) an amateur radio HF transceiver, and (H) a high band scanner. Each unit will be detailed in future chapters.*

or 160 kilometers from your location) can be quite different from another (events in your immediate area). Access to current news media along with two-way communications can easily be the most important factor in anyone's survival plan. (You do practice emergency preparedness, don't you? Even elementary school students practice fire drills!). Normal information channels, communications media, and power sources may not be available during times of crisis or emergency, and one may be forced to rely on alternate services. Anticipation and routine practice are the keys to survival. Let's begin with a look at some of the more familiar emergency situations and use them to "whet your thinking" for your own "preparedness plan."

The Need for Survival Communications

Emergency and crisis situations typically occur with little, if any, prior warning. Some are of a personal nature; others are on a large scale affecting a wide area and a great number of people. Imagine a couple traveling on an unfamiliar freeway stopping at a combination service station and convenience store, for example. The man pumps auto fuel while the spouse goes inside for snacks. But also imagine that the spouse does not return and inside attendants "never saw her." However, if a pair of micro-sized hand-held talkies were used for continuous communications, a potentially life-threatening situation might be averted.

Violent Storms with powerful lightning that can destroy homes and buildings or spark wildfires are familiar examples of emergencies. Summer storms also produce heavy rains, floods, and straight-line winds—all capable of producing severe damage and endangering lives. Being prepared with a safe shelter to "ride out" storms, plus having an alternate source of power capable of lasting 24 to 48 hours is encouraged. Monitoring NOAA weather stations, police and fire

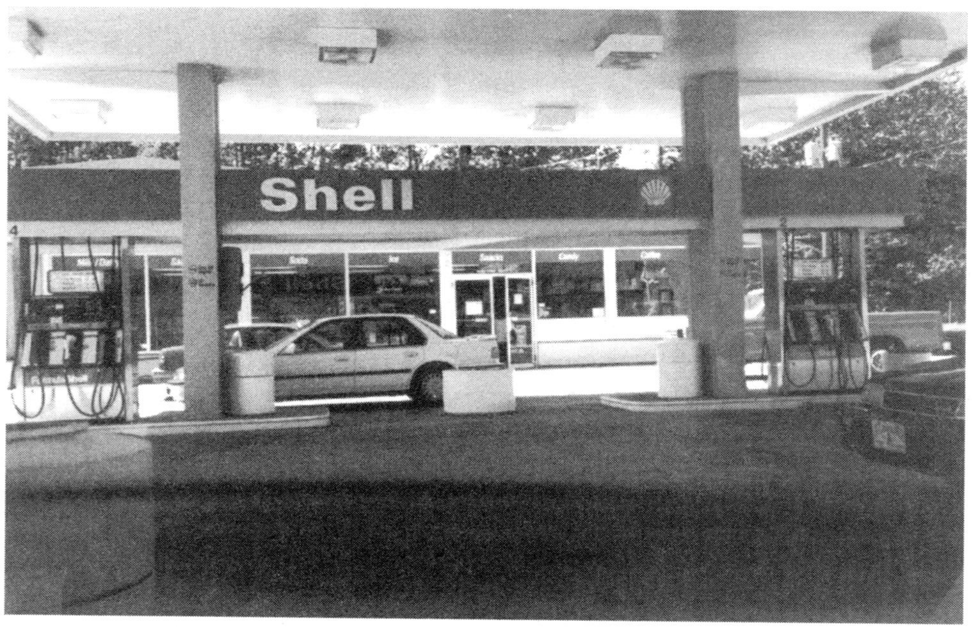

Figure 1-2 Even the most innocent looking situations hold potential for becoming serious on a moments notice. A spouse might enter a strange roadside store, for example, and not return to the auto. Inexpensive FRS talkies are ideal for staying in touch during such encounters in unfamiliar areas.

Figure 1-3 *The air near a lightning strike instantaneously rises to 50,000 degrees, producing a shock wave (thunder) and igniting objects on the ground.*

services, plus local amateur radio repeaters for up-to-the-minute "what is happening" reports is extremely beneficial. Naturally, you should have planned ahead by obtaining an up-to-date frequency list of the previously mentioned services and know how to tune them. A dependable supply of backup batteries also should be on hand and checked before needed.

Tornadoes can occur in almost any area of the United States and in almost any season of the year. They are far more devastating than strong storms, and survival preparedness includes ensuring personal safety plus at least two forms of communication. A public shelter, basement or small interior room are recommended places of refuge when tornadoes are in your area. Get there early and stay there until you know positively that it is safe to come out!

A battery-powered scanner and/or NOAA weather radio is invaluable for staying informed about "what is happening" from a place of refuge. A handheld, battery-operated transceiver is priceless for staying in touch with "the outside world" during such emergencies.

What type of handheld transceiver is recommended? The most common and familiar unit is a small CB set. Exceptionally small CB transceivers are presently being manufactured. Some even include NOAA weather reception. Their drawbacks are a limited range and interference from other CBers—especially during emergencies. Quite possibly, the best of all is an amateur radio, 2-meter/70-cm FM handheld transceiver with extended coverage of NOAA weather channels plus police and fire frequencies. The stipulation here is you must have a valid amateur radio license and an official call sign issued by the Federal Communications Commission to transmit with any amateur radio unit. **EXCEPTIONS ARE NOT ALLOWED!** Fortunately, passing the electronics theory test (plus a Morse Code proficiency test for long range communications with larger and more elaborate amateur radio transceivers) is easy. Really! Kids from 7 to 70 years of age have done it, and so can you! More details in following chapters, read on!

Wildfires also occur in any part of the country; however, they are more prevalent in

Figure 1-4 *Violent storms with accompanying lightning, hail, winds exceeding 50 m.p.h., and flash floods cause millions of dollars worth of damage and are responsible for hundreds of deaths annually.*

Figure 1-5 *Tornadoes are nature's most violent storms. They spawn hail, high straight line winds, and are deadly forces to be taken seriously. A reliable means of weather monitoring and communication is vital for staying clear of a tornado's path.*

wooded areas and during months of exceptional dryness. In today's world, entire communities have sprung up in or very close to locales susceptible to wildfires. Wildfires may be caused by lightning or a careless match. They can spread at an alarming rate, bring down power lines, and produce a health threat through smoke inhalation. Communications requirements during wildfires are quite diverse and vary according to geographic locations.

Conventional AM/FM and TV reports may be useful, but those services may be the last ones to provide pertinent information. A CB set could be useful for communications in your immediate vicinity. Bear in mind, however, a CB unit is open to blind interference by almost everyone and may provide more hearsay than facts. Having a scanner plus an accurate frequency guide that lists forestry, fire fighters, police, sheriff, amateur radio, and other agencies involved in assistance can be a superb asset during wildfires. When vacationing, camping out, or traveling as a group in areas prone to fires, personal GMRS transceivers are good for family communications in the immediate area. Finally, 2-meter and 70-centimeter amateur radio mobile and handheld transceivers are superb for two-way communications. I should also mention, both 2-meter and 70-centimeter services are included in the extended receiving range of the amateur radio 2-meter/70-cm transceivers, thus allowing one compact unit to fill many needs.

Snow and Ice Storms are well known for their ability to disrupt travel, strand motorists, isolate homes, and halt normal electrical service. Winter storms are usually preceded by warnings that wise people heed, some ignore, and others do not hear—possibly until too late for safe action. Early

Figure 1-6 *The intense heat generated by lightning often spawns wildfires which destroy thousands of acres of land each year. Knowing where such fires are located and what escape routes are open are vital for survival.*

preparation is most important for survival in blizzards and ice storms. A full discussion of this topic is far beyond the scope of this book (collecting firewood, heating fuel, storing food and water, planning trips by alternate transportation means, etc.); however, we can discuss communications needs for home, office and travel during snow and ice storms.

The basic equipment most people use for information acquisition and communication during winter storms are an AM/FM radio, TV, weather radio, and CB set—all battery powered. Assuming the storm is short-lived and outdoor travel is not necessary, those "bare-bones items" may be sufficient. Unforeseen travel and longer-than-expected storms are a different story!

Both "fresh" and spare batteries go dead after long periods of continuous use, and a CB unit is subject to interference plus range limitations. For home or office information acquisition, a VHF/UHF scanner (plus a frequency list of various public services and utilities) is highly recommended. For traveling, a VHF/UHF scanner (assuming the state in which you are traveling does not prohibit scanners in vehicles), an amateur radio 2-meter/70-cm transceiver, and an HF (low band) transceiver are almost vital. Again, I emphasize one must be licensed by the Federal Communications Commission to use "ham" equipment. No exceptions! Are there (less attractive) alternatives? Yes, a CB set, GMRS transceiver or a cell telephone. These units depend on other people/stations being nearby, but they beat sitting and doing nothing if you are stranded. Finally, a Global Positioning System (a pocket unit similar to the electronic mapping systems available in some new autos) is extremely useful for finding directions when lost in a snowstorm or in an emergency situation.

Hurricanes are also well known for their mass destruction. Fortunately, they are

Figure 1-7 *Surviving snowstorms and blizzards requires intelligent preplanning, including maintaining a stock of essentials, a reliable means of communication, and a dependable source of energy.*

Figure 1-8 The familiar units many people rely on during winter storms include an AM/FM radio, TV and a NOAA weather receiver. Their use for acquiring specific information relating to immediate needs, however, is often limited.

usually preceded by warnings and evacuation directives. The best advice available is heed such warnings before traffic jams become a nightmare and get out of the affected area early. If you cannot evacuate (such as living on a small island with its bridge out or damaged) board up windows with plywood, take shelter in the most substantial structure available, and stay there! A simple FM radio will probably be useless for information because local stations are likely to lose power. Tuning the AM band at night will allow you to receive distant stations, provided static from nearby thunderstorms is reasonably low. Scanners normally have a reception range of 50-75 miles and may not be useful during a hurricane. The same statements hold true for cellular phones unless you have one of the newer direct-satellite cell phones. Remember, however, thousands of other users may be overloading the satellite system and you may not be able to access the satellite (get a dial tone or telephone anyone). One of the best communications aids is (you guessed it) an amateur radio HF (low band) transceiver. The hitch here is it needs an outside antenna, which could be destroyed by the hurricane or a spun-off tornado. Many radio amateurs have communicated from storm shelters using indoor antennas. Their signals were not very strong, but they were the first and only signals from the areas and provided vital lifesaving communications from the devastated sections.

Earthquakes are usually considered confined to California, but they can occur anywhere from New York to Texas, in Arizona, or even Alaska. If you are caught in an earthquake, seek shelter under a sturdy table or a door frame until the ground stops shaking. Avoid standing near windows or driving across bridges or overpasses. If possible, pull to the side of the road and wait for the tremors to stop. Short range communications units like CB sets and scanners are useful for monitoring communication sources within an approximate 50-mile area of an earthquake. For reaching into and out of a stricken area from a greater distance, "low band" communications units like an amateur radio HF setup have proven extremely beneficial. Direct satellite relays for telephones and/or

Figure 1-9 Hurricanes are often called the greatest storms on earth. They spawn high winds, thunderstorms, tornadoes and floods. Evacuation also causes mass traffic jams and chaos. Communications before, during and after hurricanes play a major roll in survival.

Figure 1-10 Terrorist activities like the federal building bombing in Oklahoma City emphasize the significance of various communications systems. They also emphasize the importance of carrying a handheld talkie for communicating with others should you become a victim of an attack.

commercial communications systems may or may not be useful, depending on the number of users "loading down" those satellite systems.

Terrorist Activities including bombings, disruption of public transportation systems, and international threats have increased at an alarming rate recently. Both nuclear and chemical weapons are available on the international black market and presumably sell to the highest bidder. Many governments also operate antiterrorism counterintelligence services that exchange information internationally. Known terrorist organizations are closely monitored; however, "holes" in the system often occur. Terrorists' targets include everything from federal buildings (e.g., the Oklahoma City bombing) to airlines. They are not confined to one country or a single geographic area. Monitoring terrorist activities and threats is obviously a complex situation requiring both HF and VHF receiving equipment.

International Crises often evolve from terrorist activities and expand into political, civil, or international wars. Indeed, over a dozen world areas or small countries are typically in a state of unrest or military takeover every year. What type of accurate information acquisition is suitable for monitoring terrorist activities and international crises? A shortwave radio covering the range of 3 to 30 MHz is the best equipment for "direct from the source" information. Traditional news

A terrorist nuclear attack — is it a matter of 'when,' not 'if'?

Although a horrible tragedy, the Dhahran bombing was less than a firecracker compared to what may lie ahead in an increasingly perilous world.

A U.S. Senate committee recently studied the possibility of a terrorist attack on America, not using mere dynamite, but — hear this — a weapon of mass destruction, such as nuclear or biological. The committee's respected chairman, Sen. Sam Nunn, not given to hyperbole, concluded with these sobering words: "It is only a matter of 'when,' and not 'if.' "

Hello! Did anybody hear that? Does anyone realize what he said? The news media and everyone else hardly noticed and went right back to their obsession with cultural trivia.

Can you imagine a terrorist nuclear explosion in New York City, Washington, D.C., or Los Angeles? Welcome to the 1990s.

This is the first decade in human history when one person, or a small radical group, can obliterate an entire large city. In fact,

My Word

JIM BRAMLETT

it is not only possible but probable. Why is it now probable?

During the Cold War, the probability of nuclear mischief was kept low throughout the world by America's strategic strength, the few nuclear nations, the relative sanity on the part of leaders of the Soviet Union, their tight control over their nuclear forces, and their realization of our determination.

But the picture has radically changed since 1989, the collapse of the Soviet empire, and the loss of control over their weapons. According to intelligence reports, former Soviet nuclear weapons have found their way to Iran, the leading terrorist state, and no telling where else. Iraq, Libya and Syria are probably close to producing their own, plus chemical or biological warheads, the United Nations' feeble efforts notwithstanding. And they are driven by one compelling force: an irrational hatred of Israel and America.

The same intelligence sources report planned Iranian war games code-named Tariq al-Quds, meaning "The Road to Jerusalem," as well as Iran-backed terrorist plans to use nuclear weapons to gain this main objective, Jerusalem. This could include attempted nuclear blackmail against America or Israel, or an actual detonation in New York, Tel Aviv or elsewhere, to "send a message." Oddly, the gruesomeness of such an attack only attracts, not deters. They see more dead Western bodies as only increasing their rewards in Islamic heaven.

But neither is all quiet on the strategic front. Russia is teetering on anarchy or, more likely, a soon-to-be anti-Western, anti-Christian and anti-Semitic autocratic regime under rising star Gen. Lebed, who recently announced, "Sooner or later, victory will be mine." But under Boris Yeltsin, intelligence sources recently revealed a huge new nuclear-control center in the Ural Mountains; the Pentagon expressed concern over Russian nuclear submarines tailing our ships and lurking off our coasts; and the *Wall Street Journal* stated that Russia "is still completing construction of submarines begun in the Soviet era, on about the same timetable." Meanwhile, the United States is disarming, while sending billions of dollars to Russia.

What does all this mean? Many believe that the combined, unprecedented perils are a prelude to Armageddon, the period of time that Jesus described as a tribulation "such as was not since the beginning of the world to this time, no, nor ever shall be." (Matthew 24:21)

There are reasons to believe His words may soon come to pass.

Jim Bramlett, Orlando, is a retired lieutenant colonel in the Air Force.

'Will you accept some spiritual counseling as payment for the hit?'

Figure 1-11 *International terrorism is an ever-increasing threat. Monitoring various shortwave broadcasts and services around the world allows one to study the facts with a minimum of bias and sensationalism.*

media, such as radio and television, are often biased, slanted, or sensationalized, thus the public is subjected to what broadcasters want them to hear rather than the full story. Likewise, newspapers often report "slanted" views and facts based on political directives. Tuning in long-established authorities like the Voice of America, British Broadcasting Corporation (BBC), Deutsche Welle (German newscast), and other established networks often provide additional views. Combined with shortwave broadcasts directly from affected or influencing areas (e.g., Radio Baghdad, Radio South Africa, etc.), listeners may derive their own conclusions. An additional chapter is devoted entirely to shortwave listening and readers are encouraged to consider it vital reading.

Our list of needs for survival communications could continue for many more pages, but its main points are obvious. Equipment for monitoring activities and communicating with others during any type of emergency or abnormal situation is a priceless asset. Some readers may feel a need to prepare for only one or two of our previously discussed situations. Others may want to gear-up for any and all unusual events. Hopefully, this introduction provided basic insight into communications equipment applicable to your particular lifestyle and needs.

Keynotes to Survival: Knowledge and Preparedness

Left isolated and "in the dark" in today's world is a scary situation. Yes, and even if a myriad of communication's assets are at your fingertips, they are of little benefit unless you are familiar with their applications and have practiced operating them before they are

Figure 1-12 *The main operating console of Deutsche Welle, the voice of Germany. The shortwave station is well known and respected for its accurate news reports.*

needed. Attempting to figure out how to operate a complex radio or tune a specific frequency (or discovering the radio has dead batteries!) precisely when they are needed can only be described as more frustrating than helpful. Never underestimate how rapidly emergencies or the urgent need for immediate communications can arise during such times. As examples, ask yourself what you would do if someone attempted to kidnap you or carjack your auto while you were stopped at a traffic light right now. Visualize how you would handle communications' needs if you were traveling during the winter and suddenly ran into a snowstorm or blizzard. What information sources would you tune in if local radio and TV stations broke in with an announcement that a world crisis had just erupted in the Far East? Would those sources be accurate or prone to sensationalism or propaganda angles? Remember the basic purpose of terrorism is to disrupt social structures, produce chaos, confusion and distrust.

In the same way different supplies are necessary to combat various types of weather-related emergencies, different types of communication equipment are required for facing other emergency situations. Is there one deluxe unit that would fit all needs? Not really. Further, such a physically large and expensive communications unit would not be capable of monitoring a variety of sources simultaneously. A simple failure in one of its basic circuits (such as a power supply or audio amplifier) could render the unit completely inoperative. Separate radios, scanners, handhelds and transceivers have their advantages!

Personal and Local Area Communications Media

A large number of communications media and commercial two-way radios are at your disposal. Let us briefly discuss the more popular ones in a "what is it and how you use it manner."

The Citizen Band makes a good starting point. Almost everyone has heard of CB radios or seen them in use. During the early days of CB, an F.C.C. license was required for operation. Those licensing stipulations were dropped several years ago, which means you can purchase a CB two-way radio and begin using it immediately. There are three basic classes of CB sets: mobile units, home/base units, and compact handheld transceivers. The inexpensive varieties of each usually operate in the AM mode and are good for talking with others on any of the 40 CB channels. "Handles" are normally used for names and station identification. The more sophisticated home/base CB units operate in both AM and SSB modes and some are quite elaborate.

Although single sideband (SSB) usually exhibits better range than AM, only a small number of people use it. CB channel 9 has been set aside for emergency communications and traveling assistance. Members of REACT (Radio Emergency Associated Communications Teams) throughout the United States monitor Channel 9 on a continuous basis. In a typical year, REACT handles over 100,000 requests for assistance. They range from assisting stalled vehicles to life-threatening emergencies. That is a quite impressive record for an unlicensed service available to everyone.

Are you involved in boating activities? The VHF marine band is comprised of many FM channels between 156 and 162 MHz, and is used primarily for short-range, ship-to-ship and ship-to-shore communications. Most important is channel 16 on 156.800 MHz, which is a distress, safety and calling channel. Both mobile and handheld transceivers are available for marine two-way radio operation. They typically have a range of 5 to 25 miles. Marine channels follow a specific band plan and should be used in exact accordance with the plan. Information on channels and operation are usually included with the unit. A license is required for marine radio operation, and instructions on applying for the license are usually packed with the transceiver. Alternately, one can obtain a license application by writing the Federal Communication Commission, 1270 Fairfield Road, Gettysburg, PA 17325-7245.

Although some groups and commercial services use marine radios for inland operation, the concept is really not ethical. An official business radio service should be used instead. Business radios are available in both mobile and handheld versions, and typically have an operating range of 25 miles. Repeater stations are often used to extend the range. Business radio operations are in several bands: the most popular are 150-173 MHz and 457-470 MHz. A relatively new "T" band has evolved between 470 and 512 MHz. Some familiar services using commercial two-way radios are police, fire, local power and gas companies, and even local radio and TV stations.

The General Mobile Radio Service (GMRS) band is located in the 462-MHz range, and could be viewed as a new evolution in CB. It was originally intended for use by quasi-professional services like shopping mall security forces, large churches directing traffic in their parking area, etc. GMRS also has a REACT affiliated channel for citizens involved in search and rescue activities like missing persons, etc. The more "professional air" of GMRS makes it quite attractive for nontechnical minded groups needing short range communications on an impromptu basis. GMRS transceivers are available nationwide, the cost is low, and an application for the required F.C.C. license is usually packed with each new unit.

A recently authorized Family Radio Service (FRS) is also based in the 462-MHz range and activities are conducted on frequencies in between normal GMRS channels. FRS is designated as a license-free and low-power communications medium for families to use as needed. Pocket-size talkies available from several manufacturers are addressing this area, and it promises to be a highly effective resource for both emergency preparedness and daily communications within a one to five or ten mile area. Additional information on FRS is included in the chapter on Personal Communications Systems.

Cellular telephones operate in the 850- to 890-MHz range, usually have a communications distance of 10 miles, and depend on "cells" or repeater stations at various locations for connections to telephone lines. They are a good medium provided cells are in range and one can afford their cost-per-minute air time charges.

Amateur Radio 2-meter/144-MHz and 70-cm/440-MHz transceivers are extremely good for communications over an approximate 50 mile area almost anywhere in the United States. That is because there are numerous 2-meter and 70-cm repeater stations located throughout the country. As mentioned earlier, the prime consideration with any amateur radio equipment is first obtaining a valid license from the Federal Communications Commission. A formal examination involving simple electronic theory and equipment operation is required for the basic license that permits operation on the popular 2-meter and 70-cm bands. Probably the three most popular frequencies in the United States for monitoring are 146.94, 146.88, and 146.76 MHz. Additional licenses, which include both advanced electronic theory and Morse code proficiency, are required for operation on lower frequencies and longer range bands. Even if you do not have an amateur radio license at this time, tuning in their activities on repeater channels is encouraged during both normal times and times of emergency or crisis. They have proven to be a prime source for "what is happening" information in many areas of the USA.

Accurate Information Vital in Situations of Urgency

Whether seeking details on an impending snowstorm, "inside facts" regarding a nationwide terrorist activity, or information on an international problem, accuracy is always the most important consideration. Are all media reports similar and accurate? Maybe, but do not kid yourself or be lulled into accepting media reports at face value. Most reports are influenced toward sensationalism, and plain old propaganda. How can you ensure a high degree of accuracy? By checking several different information sources and using that breadth of information to derive a logical conclusion. Remember that any unusual type of occurrence, especially one involving loss of normal power and telephone service, catches people off guard and inspires opinionated chatter. These statements hold true for both printed and broadcast matters, and include everything from newspapers and regular radio/TV broadcasts to CB and commercial shortwave stations.

Localized Crises like snowstorms, tornadoes or emergencies in your immediate area are usually easy to check for accuracy. Tuning in NOAA weather stations or main police and fire repeaters on a scanner should reveal the facts. Likewise, monitoring our previously mentioned "main" amateur radio 2-meter repeaters should add additional insight (amateur radio setups are probably at weather bureaus, police departments, and civil defense locations, which means you hear the information firsthand). If you are traveling or are in a more isolated area, the CB radio may be the only mass communications facility in your range. Remember CB is good for looking ahead or behind you in traffic, but may be susceptible to much hearsay in other instances. If you are near a coastline, try scanning the VHF marine channels in the 157- to 162-MHz range for views from near-shore mariners. Some people may ask if on-line computer services like the Internet are useful (assuming one has power or a battery-operated laptop computer). Quite frankly, they are about the last place to turn for information. Not only is the "Information Super Highway" wide open to opinions from anyone, it is quite time-consuming, opinionated and of questionable accuracy.

National Events such as terrorists bombings, political unrest or inquiring about people in distant areas affected by local crises usually requires communications/monitoring equipment with a slightly greater broadcast/reception range. A common, yet often overlooked, item for use here is an AM radio. During evening hours, signals from the smallest radio stations throughout the United States "skip" off the ionosphere and can be received around the country. These stations are numerous, which means listing all of them and their frequencies would be a book in itself, so a general "tuning for news" concept is logical. Even if you do not find a station in a desired/ affected area within the first few minutes of listening, you will probably hear a variety of other reports on the situation from various stations around the country. An ideal time for monitoring is around dusk or dawn as well as two minutes before each hour. AM stations usually identify themselves, giving their call letters and location on the hour, and then many go right into a news broadcast. Another good source of information is commercial shortwave stations in the United States broadcasting on international bands. As an example, the Voice of America is located in Pennsylvania, WRNO worldwide in Louisiana, WWCR in Tennessee, and WEWN in Alabama. The last two stations are religious oriented and may offer additional insight not available through usual media. Low band amateur radio ("ham bands" between 1.8 and 30 MHz) is another quite reliable area for information acquisition concerning national events. Usually, a portable amateur radio station is set up in an affected area and anyone with a shortwave receiver can monitor related communications. Particular frequency ranges to tune during morning and early evening hours are 3,800 to 4,000 kHz, 7200 to 7400 kHz, 14,200 and 14,350 kHz. During late night hours, the two lower frequency ranges are suggested. A few words regarding signal propagation and frequencies are appropriate at this time. Upper frequency bands are good for daytime long-range use and lower frequency bands are good for long-range nighttime use. For in-country use, however, more "middle of the spectrum" ranges are good for both day and night. (See Figure 1-13 and 1-14.) More details on signal propagation will be presented in our upcoming chapters on shortwave and amateur radio.

Figure 1-13 *Explosions on the sun's surface are called sun spots, and they directly influence long range communications in the 3-30 MHz range. The greater the sun spot count, the better the long distance communications on HF frequencies.*

International Situations affect all sectors of our society, because everyone is concerned with happenings in today's world. One medium truly does not serve all needs in this respect. Radio and TV reports are only one viewpoint, and shortwave broadcast stations often feature eye-opening reports relative to situations of the day. Here, we would probably listen to the evening broadcasts, because they are typically beamed into the U.S. from afar. We must realize classic propaganda did not become extinct after the era of Tokyo Rose during WWII. When England invaded the Falkland Islands a few years ago, for example, "Argentina Annie" took to the airwaves on

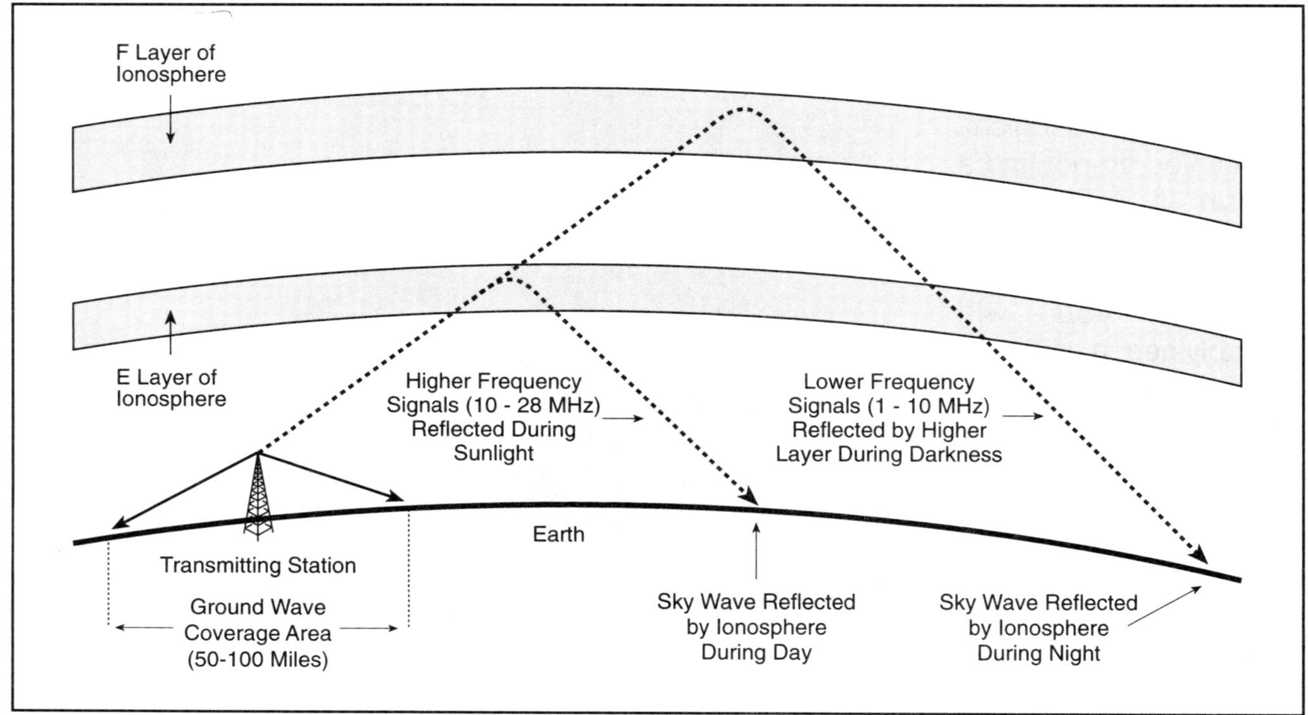

Figure 1-14 *Photon energy from the sun causes heating and cooling of the earth's ionospheric layers, resulting in an "electrical mirror" for reflecting radio signals long distances. Frequencies between 10 and 28 MHz reflect off lower layers during the day, frequencies between 1.0 and 10 MHz reflect off higher layers during the night.*

Radio Argentina. The station gave reports of its country's status and situations of the Falkland Islands, then played British rock music for enroute Marine troops and "Argentina Annie" emceed with segments of "Where are you going, Tommy?" During Desert Storm several years ago, "Baghdad Bennie" was an evening personality on Baghdad radio. The saga continues today, and probably will continue forever! Stations such as WWCR, WEWN, and Radio Australia have proven beneficial for separating fact from fiction infiltrating international shortwave broadcasts.

Overall, I suggest using VHF/UHF monitoring equipment for details in localized situations, a shortwave radio and a good AM radio for monitoring national events, and a shortwave radio capable of tuning both amateur radio and international shortwave bands for situations of global concern.

Overview of The Communications Spectrum

Up to this point, we have discussed various communication services in a nontechnical manner. We have also mentioned frequency ranges of each, using familiar terms of kilohertz (kHz) and megahertz (MHz). Now let's shift focus and survey the full electromagnetic spectrum and visualize where many of the popular services are located. Hopefully, this will give you a good working idea of their coverage area and susceptibility to interference. Please refer to Figure 1-15 as we continue.

Radio waves, which include television waves, microwaves, signals of all types, plus actual light waves are measured in meters or centimeters according to the actual length of their individual waves and are commonly referred to in kHz, MHz, GHz and THz. The

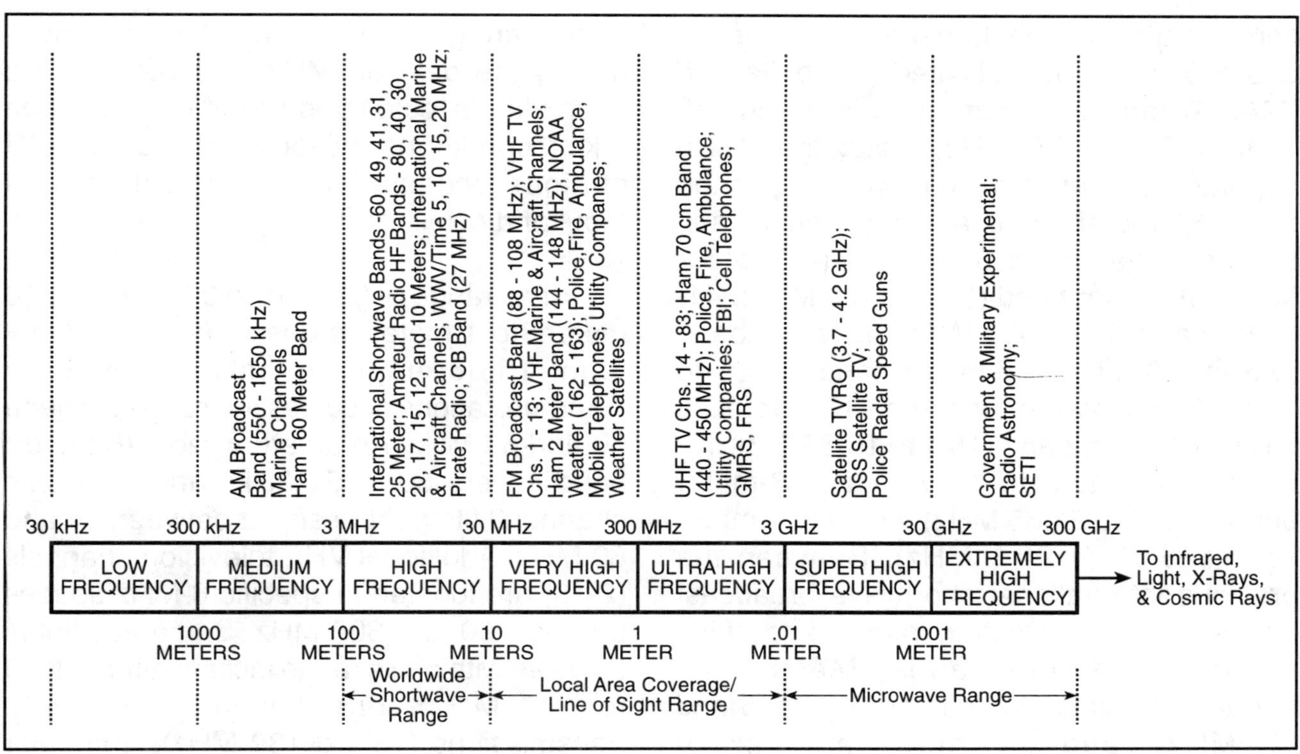

Figure 1-15 *A condensed outline of the radio frequency spectrum. Frequencies between 500 kHz and 30 MHz reflect or "skip" off the earth's ionosphere for long range communications. Frequencies above 30 MHz travel in a straight line-of-sight path. Discussion in text.*

shorter the wavelength (in meters), the higher the frequency (in MHz). As an opening point of reference, regular sound (audible) waves are in the audio frequency (AF) range of 20 Hz to 20 kHz. Normal conversation is within the 750- to 5,000-Hz range. Female voices are usually in the 1,000- to 5,000-Hz range. Conventional AM radio conveys audio frequencies between 300 and 5,000 Hz for music. FM stereo stations convey audio frequencies between 300 Hz and 20,000 Hz. Moving above audio frequencies, we enter the radio frequency (RF) spectrum in the low frequency range of 30 kHz to 300 kHz. Small two-way communicators use these ultra-low frequencies. This range is limited because the wavelengths are quite long and related antennas would require several hundred feet of wire. Next is the medium frequency range of 300 kHz to 3 MHz (3,000 kHz). Near the lower end of this range is the standard AM broadcast band on 550 to 1650 kHz. Between 1700 and 3000 kHz, there is an amateur radio 160-meter band and several marine mobile single sideband (SSB) channels. There is also an international marine and distress frequency on 2182 kHz. During the day, stations using MF (medium frequency) bands/ frequencies typically have a coverage area or communications range between 50 and 150 miles. During the evening and night, their coverage/range extends out to 1000, 2000, 4000 miles, or possibly worldwide, depending on ionospheric conditions and power levels.

The range of 3 MHz to 30 MHz (which, incidentally, corresponds to wavelengths of 100 meters to 10 meters) comprises the HF or high frequency part of the spectrum. This is the hottest area of long distance (worldwide!) communications in the radio spectrum. Between 3 and 4 MHz, there are some marine channels used for high seas operations, Military Affiliated Radio Service (MARS), and an amateur radio band (80 meters: 3.5 to 4.0 MHz). Also, there is a popular 120-meter tropical broadcast band below 80 meters, plus additional shortwave broadcast bands on 60 meters (4.6 to 5.1 MHz), 49-meter band (5.95 to 6.2 MHz), 41-meter band (7.1 to 7.3 MHz), 31-meter band (9.5 to 9.9 MHz), 25-meter band (11.65 to 12.05 MHz), 22-meter band (13.6 to 13.8 MHz), 19-meter band (15.1 to 15.6 MHz), 16-meter band (17.5 to 17.9 MHz), 13-meter band (21.45 to 21.85 MHz), and an 11-meter band (25.67 to 26.1 MHz). Between the previous international shortwave broadcast bands are eight amateur radio bands plus numerous marine bands, MARS/CAP frequencies, time services on 5, 10, 15 and 20 MHz), and an interesting mix of underground, pirate and mysterious "guerilla" activities—all of which can be tuned with a shortwave receiver. At the high end of this range (approximately 27 MHz) is the popular Citizens Band. Generally speaking, frequencies between 100 meters and 30 meters (approximately 10 MHz) are good for covering long distances during the night and frequencies between 30 meters and 10 meters are good for covering long distances during the day. 30 MHz/10 meters is the approximate "cutoff frequency" for long range "skip" conditions, which explains why CB stations occasionally reach out several thousand miles.

The range of 30 MHz to 300 MHz, or 10 meters to 1 meter, is designated VHF (very high frequency). A mix of military communications occupies the range between 30 and 50 MHz. An amateur radio (6 meter) band covers 50 to 54 MHz, and television channel 2 fits right beside it between 54 and 60 MHz. Additional VHF television channels (2-13) are located in specific 6-MHz spaces between 60 and 300 MHz. Some additional services within this range include aircraft (119 to 122 MHz), direct weather satellite transmissions (137 to 139 MHz), a popular amateur radio "ham" band (144 to 148 MHz), and the NOAA weather radio band (162 to 163 MHz). Also in this range are the VHF marine channels and telephone services (157.2 to 157.425) plus a variety of business radio related services (153 to 174 MHz). The latter channels are often used by police, fire, ambulance, utility companies, hospital rescue helicopters, etc. Rural and smaller city police

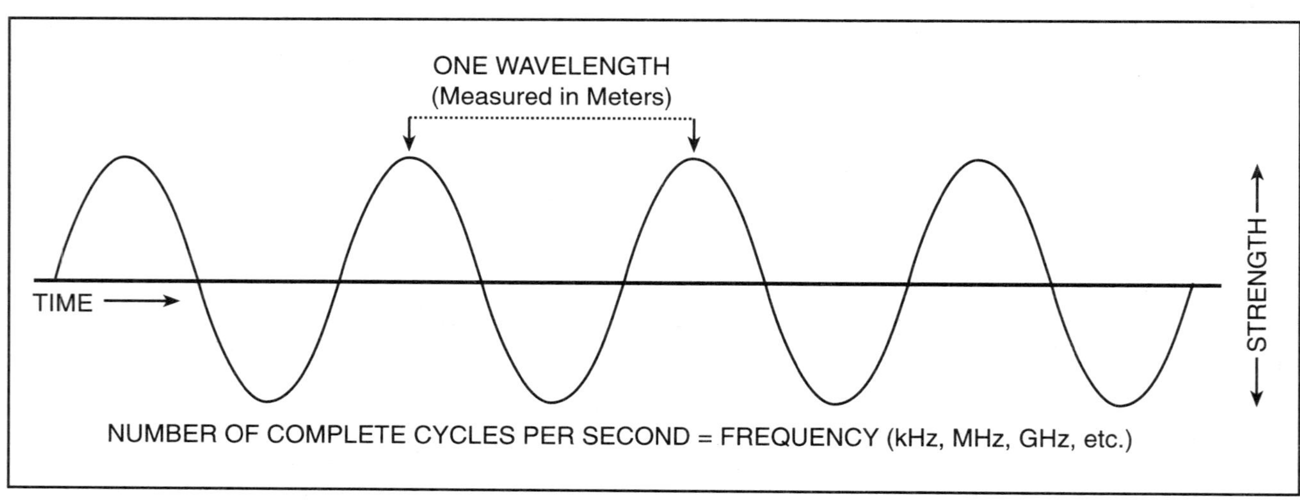

Figure 1-16 *The frequency of radio waves is referred to in number of cycles per second, and measured in meters or centimeters of each wave's length. Higher frequencies correspond to shorter wavelengths.*

Frequency in Kilohertz (kHz)	Frequency in Megahertz (MHz)	Meter Band
3800 - 4000	3.9 - 4.0	75
4600 - 5100	4.6 - 5.1	60
5950 - 6200	5.95 - 6.20	49
7100 - 7300	7.1 - 7.3	41
9500 - 9900	9.5 - 9.9	31
11650 - 12050	11.65 - 12.05	25
13600 - 13800	13.6 - 13.8	22
15100 - 15600	15.1 - 15.6	19
17550 - 17900	17.5 - 17.9	16
21450 - 21850	21.45 - 21.85	13
25670 - 26100	25.67 - 26.10	11

Figure 1-17 *Popular shortwave bands used for international broadcasts. Also shown for frequency referencing is the Amateur Radio 75-meter band.*

forces are found in this range; however, larger metropolitan police services have migrated into the UHF range of 450 to 500 MHz.

The 300-MHz to 3-GHz (3000 MHz) range is designated UHF for "ultra high" frequency. Within this part of the spectrum are television channels 14 through 83 (scattered between 512 and 806 MHz), four UHF amateur radio bands, UHF police and utility services (450 to 470 MHz), the "T" band used by sheriffs, detectives, FBI, etc. (470 to 512 MHz) and cellular telephone services (between 850 and 900 MHz). The new cordless telephones used in homes are on 900 to 905 MHz

The range of 3 GHz to 30 GHz is SHF, or super high frequency. This range is used for the well-known "TVRO" satellite dishes that provide home TV reception, and the DSS direct broadcast satellite (The 18-inch dish systems), GPS (global positioning systems), and a variety of military activities. I should also mention microwaves start at 2 GHz and continue to slightly beyond 300 GHz. Most home microwave ovens operate in the 2- to 4-GHz range; however, they transmit into a shielded internal "cavity" to heat food rather than transmit a signal into the air.

Few communication services occupy frequencies above 300 GHz, but I think you will find the following notes interesting. If we continue upward in frequency to 1000 GHz it will become 1 Terahertz (THz). We are now approaching the frequency of infrared light. If we continue to almost 1000 THz (referred to as Petahertz PHz), we go into the visible light spectrum. The color red is lowest in frequency, green is in the approximate mid-range of our eyes, and violet is toward the upper frequency limit of our eyes. Continuing beyond 1 PHz we come to the ultraviolet region, then x-rays, gamma rays and cosmic rays. We have now completed the overview of the spectrum from sound through light.

Let's now add some interesting side notes to the previous discussions. First, decisions regarding what services may use various frequencies in the electromagnetic spectrum are determined at World Radio Conferences in Geneva, Switzerland, every few years. There are over 300 countries in the world and each country, regardless of size, is granted one vote. Because frequencies between 3 MHz and 30 MHz "skip" off the earth's ionosphere, they are considered prime spectrum. Frequencies above 30 MHz are primarily line-of-sight, which means they can be "reused" by various services approximately 200 miles apart without interference. Some of you are thinking VHF and UHF TV reports seemingly transcend the line-of-sight restriction rather than being confined by it. However, their transmissions are being relayed by satellites high above the earth. From their vantage point 20,000 miles in space, satellites can "see" approximately one-third of the world's surface. It is also line-of-sight to the moon and stars, thus UHF and microwave frequencies are ideal for radio astronomy and deep space probes.

Another interesting point worthy of consideration is a signal's power level. In the case of audio frequencies, power is usually measured in decibels (dBs). 30 dB is equivalent to a quiet room, 60 dB equivalent to normal conversation, 90 dB equivalent to shouting or a jackhammer, and 120 dB is the threshold of pain. Moving into the RF spectrum, power levels of 100 watts to 1000 watts are capable of reaching around the world on frequencies below 30 MHz. Many stations run less than 100 watts, others run more than 1000 watts in commercial broadcast facilities. Between 30 MHz and 300 MHz, power levels between 1 watt and 100 watts are common. Between 300 MHz and 900 MHz, power levels are typically between 1 watt and 10 watts. Above 900 MHz, lower power levels are usually employed when radiated signals can affect the population at large.

Finally, I should point out that electronic designs in equipment change quite noticeably between HF and VHF/UHF frequencies and again on frequencies above 2 GHz. That is one of the main reasons why a single receiver or transceiver covering all frequencies is expensive: It is like three units (or more) in a single box! Please understand the previous discussion pointed out only some of the more popular services and activities within the electromagnetic spectrum. Including all of them simply would be impossible!

Legal Stipulations and Licensing Considerations

Although a license is not required for merely listening to transmissions and broadcasts on any frequency, a certain amount of understanding and responsibility are involved in general monitoring. This section provides a brief overview of those considerations.

Certain provisions within the communications act of 1934 (47 USC 605) as well as regulations in the International Telecommunications Union make it illegal to repeat contents of most two-way communications (except those associated with amateur radio and the Citizens Band). The communications privacy act plus various laws of individual states may also apply to what individuals may monitor and the type of equipment used (some states, for example, prohibit in-car scanners). It is the responsibility of each listener to be aware of such rules and regulations applicable to their situation.

Additionally, Congress passed the somewhat controversial Electronic Communications Privacy Act (ECPA, 18 USC 2510-2511) in 1986. This ECPA Act prohibits the public at large from listening to cellular telephone or cordless phone calls, ship-to-shore phone calls, and scrambled or encrypted satellite transmissions.

With the exception of the Citizens Band, the new Family Radio Service and/or hobbyist-type 49-MHz "walkie-talkies," requires an F.C.C. license before transmitting with a two-way transceiver. A synopsis of licensing requirements follows.

Land Mobile license applications require F.C.C. Form 155. The application fee is usually $35.00 for each associated vehicle. There may also be an operator's fee, depending on proposed rule-making. Assuming the transceiver is used on a business band, it should also be assigned a frequency before operation. Contact NABER at 800-759-0300 for details.

Business Radio Service licensing requires submitting an application on F.C.C. Form 574 for mobile, portable and/or base stations. The associated fee is $35.00. A business radio must also be assigned a frequency before use, and frequency searching/coordinating is handled through NABER 800-759- 0300 (fee required).

Maritime Radio Service on VHF/UHF frequencies use F.C.C. Form 506 when applying for a portable/mobile license, and Form 503 when applying for a base station license for contacting mobiles in near-shore/ domestic waters. Shoreside operators should also apply for a restricted operator's permit using Form 753. Applicable fees are $35.00.

International Communication Stations aboard ships cruising international waters and using HF bands must apply for a ship station license and a restricted radio telephone operator's permit, both via Form 753. A filing fee of $35.00 is required.

Please note the previous high seas license is only for possessing and operating two-way radio equipment. A valid commercial radio license is required for repairing and making internal adjustments on those transceivers. Extensive study and electronics knowledge is required to pass the exam for a commercial radio license. Vessels traveling internationally cannot chance equipment failures at sea, consequently, a commercially-licensed operator/technician "radio man" is usually a member of the crew.

General Mobile Radio Service (GMRS) does not require a radio operator's license, but does require a valid equipment license. It is applied for on F.C.C. Form 574 (which, incidentally, is usually packed with each new GMRS transceiver sold).

Amateur Radio requires a license application on F.C.C. Form 610. Several "classes" of amateur radio license are presently available and all require some knowledge of electronic theory and two-way communications practices. Amateur radio's introductory license requires only a basic knowledge of electronic technology, and kids to adults have successfully passed the exam. Operating privileges are confined to amateur radio bands above 50 MHz. Additional amateur radio licenses for operating lower frequency and longer range bands require demonstrating proficiency in Morse code copy plus passing written exams of various knowledge levels of electronics . Again, even the most difficult amateur radio license is relatively easy to obtain with a little dedication and study (even your author did it while in elementary school). Amateur radio license classes and exams are presently handled through radio clubs and volunteer examiner groups in almost every city throughout the U.S. At present, a fee of $6.00 to cover general exam administration is usually involved.

For additional information on license applications and forms, you may contact the Federal Communications Commission form office at 202-632-3676. Please remember that two-way transceivers, including low-power handheld units, must be licensed before use. Do not press any mike button or transmit until the license is in hand. Failure to obtain a proper license could result in an F.C.C. citation and lead to a fine. Again, we remind you a license is not required for listening to most shortwave services, operating a CB radio or using a handheld 49-MHz unit.

As this opening chapter pointed out, survival communications involves bringing together a number of areas and a variety of associated equipment for personal security and emergency preparedness. All areas are not necessarily applicable to all lifestyles and needs at the same time, but knowledge of available options is your advantage when facing each challenge as it arises.

Keeping those facts in mind, the following chapters zero in on specific areas of survival communications and provide guidance for using related equipment. This arrangement allows you to address both present and future needs in the most logical and cost-effective manner possible. Read on!

CHAPTER TWO

Traditional Sources of News and Information

Unquestionably, the most common source of news and information during emergencies or periods of uncertainty is a regular AM/FM radio or TV. Why so? Because they are plentiful, inexpensive and we have been conditioned to using them since childhood. Radios are in nearly every vehicle; they are sold in grocery stores, variety stores, and truck stops throughout the country. Flipping a radio on to check weather conditions or review daily occurrences is a natural part of our lifestyle. General media reports on local happenings are usually accurate (and easily cross-checked for legitimacy). Matters of national or international concern, however, are susceptible to political influence or network bias. One recent example relates to the destruction of TWA Flight 800 off the coast of Long Island, New York in July 1996. The president stated new and additional security measures aimed at protecting citizens would be implemented immediately and the new level of sophisticated equipment could increase airfares. Independent stations in many areas of the country excerpted segments of that network speech for use in their local newscasts. The one I remember most vividly began not with the latest update on Flight 800's crash but with the statement: "The president said airfares may soon increase." The announcement was followed by an edited part of the president's speech in which he said, ". . . this may increase the cost of airfares." The statement was true, but "separating the wheat from the chaff" required intelligent thinking and more information. Additional examples are endless and subjects taken out of context by the media vary from medicine to automobiles to Desert Storm involvements and much more. Misinformation is created when only part of a story is

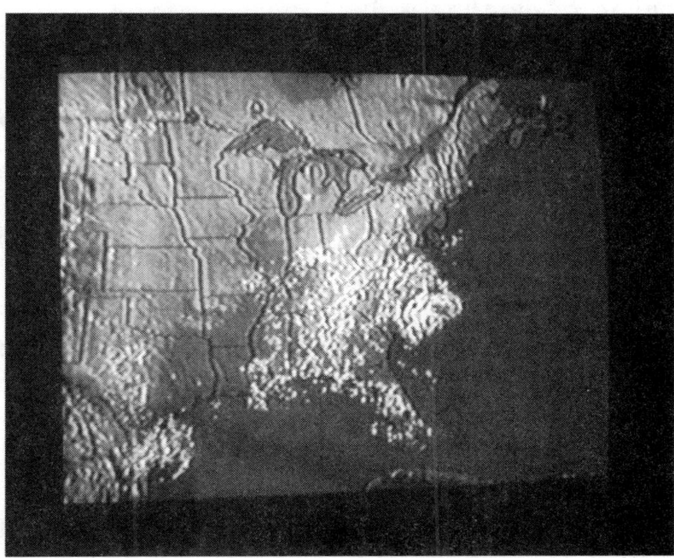

Figure 2-1 *Tuning in weather reports from local TV stations is one of the most common activities in usual survival thinking. Fortunately, weather reports are usually accurate.*

U.S. nearly faced germ war

Documents: Iraq had 'cancer' toxin

By Dave Parks
News staff writer

Iraq went to the brink of launching biological weapons during the Persian Gulf War, and its arsenal included a mysterious, cancer-causing toxin, according to secret documents now coming to light.

Many of the documents were declassified recently by the Central Intelligence Agency as part of its search for clues to baffling illnesses linked to the Persian Gulf War.

More than 20,000 troops returned from the 1991 conflict suffering from difficult-to-diagnose ailments such as aching joints, fatigue and memory loss. Also, there is growing concern that cancer rates are rising among Desert Storm veterans.

The CIA documents were made public through GulfLINK, a Department of Defense site on the Internet. The information was accompanied by a warning that it came from a variety of sources, including some not assessed for reliability.

The CIA has concluded there is no evidence that Iraq used biological agents in the gulf war, but some documents challenge such a cut-and-dry assessment.

One of the most dramatic documents was a CIA report distributed in 1992. It details an unsuccessful attempt by Iraq early in the war to launch a biological attack using Soviet-made jets.

"The plan called for a test mission of three MIG-21's to conduct an air raid ... using conventional high explosive ordnance," the report said. "Then a second mission was to take off within a few days of the first, using the same flight path and approaches."

The second mission of another three MIG-21's was to provide cover for a low-flying SU-22 aircraft carrying a

See **Iraq,** Page **10A**

Figure 2-2 *Traditional sources of news and information are a natural part of our daily lives, but their reports are often subject to bias and sensationalism. Seeking alternate opinions to confirm or refute the accuracy of statements is usually quite beneficial.*

broadcast, either intentionally or unintentionally. I am sure readers can relate to that statement!

Seeking out second (and third!) opinions on media reports can often prove quite fruitful and eye-opening. That is the main purpose of this chapter. Bear in mind that "survival communications" is also comprised of many avenues. Not only is it acquiring additional opinions on news items, it is also collecting information on events out of our immediate area, staying abreast of general information while traveling, maintaining a vital link with the outside world for citizens in isolated areas, and much more.

Major Radio and TV Networks

Most people are familiar with the traditional networks of NBC (National Broadcasting Corporation), CBS (Columbia Broadcasting System), ABC (American Broadcasting Corporation), Mutual, and CNN (Cable News Network). A number of additional independent networks also disseminate news and information throughout the United States. The purpose of the news departments of these networks is twofold: informing the public and producing high viewer ratings (which equates to high revenue). As a result, they are subject to sensationalism and political influence. They are also susceptible to every other scheme known to modern man for twisting truths in various ways. These facts are mentioned here not to devalue the vitally important role the networks play in our society, but to make you aware that news from a single viewpoint is simply that and should not be accepted as the absolute authoritative word.

Rediscovering AM (Standard Broadcast) Radio

If you were born around 1960, there is a fair probability you grew up listening to the FM broadcast band (88-108 MHz) and assumed the AM broadcast band (530 kHz - 1650 kHz) was a vague wasteland—possibly with less range than FM. My sentiments go out to individuals leading such sheltered lives. The RF spectrum above 50 MHz is less premium, so FM stations can use a wider bandwidth and transmit higher fidelity sound than an AM station. During the day both AM and FM have the same approximate range.

During the night, signals from lower frequency AM stations "skip" off the earth's ionosphere and are easily received throughout North America. Fancy or expensive receivers are not required for receiving these signals—your kitchen or car radio is quite adequate for tuning in both local and distant stations. Really! Try it! There is a fascinating world of listening on AM at night—particularly when you tune in "super stations" on wide-open or "clear" channels. Such stations provide everything from specific interest entertainment to travelers' information. As a guide to AM band tuning, check out the list of clear channel stations shown in Figure 2-3. Some of the most notable stations in history are still active on AM; they operate 24 hours a day and typically run the maximum power allowable, 50,000 watts. Local-area AM stations often occupy the same frequencies as "super stations" during the day; however, they sign off at dusk and return to the air at dawn. This makes the clear-channel stations "smooth copy" from almost any location. As you will notice in our clear channel list, some frequencies are occupied by more than one station. Which one you receive will depend on your location and the signal propagation characteristics of the season and day. One evening you may receive a station from one side of the country; another evening you may receive a different station on the same frequency from the opposite side of the country.

Looking at our clear frequency chart, your attention is directed to WWVA in Wheeling, West Virginia (1170 kHz) This station is well known for its country music, weather/traveling information and reports on interstate highway conditions. It is a quite popular station among truckers. Indeed, this one may be more enticing than CB channel 19 during late night hours! People call WWVA from all over the country, and some of the commentary is fascinating. While writing this chapter, I tuned to 1170 and signal propagation was favoring KVOO in Tulsa, Oklahoma rather than WWVA. The format was similar. Although I only listened for a few

CLEAR CHANNEL AM RADIO STATIONS

FREQ.	STATION
540	CBK, Regina, Saskatchewan
640	KFI Los Angeles, California
650	WSM Nashville,Tennessee
660	WFAN New York, New York
670	WMAQ Chicago, Illinois
680	WRKO Boston, Massachusetts
690	CBF Montreal, Quebec
700	WLW Cincinnati, Ohio
710	WOR New York, New York
	KIRO Seattle, Washington
720	WGN Chicago, Illinois
730	CKLG Vancouver, British Columbia
740	KCBS San Francisco, California
750	WSB Atlanta, Georgia
760	WJR Detroit, Michigan
770	WABC New York, New York

FREQ.	STATION
780	WBBM Chicago, Illinois
800	CKLW Windsor, Ontario
810	WGY Schenectady, New York
820	WBAP Fort Worth, Texas
830	WCCO Minneapolis, Minnesota
840	WHAS Louisville, Kentucky
850	KOA Denver, Colorado
	WHDH Boston, Massachusetts
870	WWL New Orleans, Louisiana
880	WCBS New York, New York
890	WLS Chicago, Illinois
940	KFRE Fresno, California
1000	WLUP Chicago, Illinois
	KOMO Seattle, Washington
1010	WINS New York, New York
1020	KDKA Pittsburgh, Pennsylvania

(continued)

CLEAR CHANNEL AM RADIO STATIONS (CONTINUED)

FREQ.	STATION	FREQ.	STATION
1030	WBZ Boston, Massachusetts	1180	RADIO MARTI Marathon Key, Florida
1040	WHO Des Moines, Iowa	1190	KEX Portland, Oregon
1050	WEVD New York, New York	1200	WOAI San Antonio, Texas
1060	KYW Philadelphia, Pennsylvania	1210	WOGL Philadelphia, Pennsylvania
1070	KNX Los Angeles, California	1220	WKNR Cleveland, Ohio
1080	WTIC Hartford, Connecticut	1500	WTOP Washington, DC
1090	WEAL Baltimore, Maryland	1510	WLAC Nashville, Tennessee
	KING Seattle, Washington	1520	KOMA Oklahoma City, Oklahoma
1100	WWWE Cleveland, Ohio	1530	WCKY Cincinnati, Ohio
1120	KMOX St. Louis, Missouri	1540	KXEL Waterloo, Iowa
1130	WBBR New York, New York	1560	WQEW New York, New York
1140	WRVA Richmond, Virginia	1570	CKLM Laval, Quebec
1160	KSL Salt Lake City, Utah	1580	KBLA Santa Monica, California
1170	WWVA Wheeling, West Virginia		
	KVOO Tulsa, Oklahoma		

Figure 2-3 *A condensed listing of clear channel AM stations transmitting in the 530- to 1650-kHz range. Many of these stations can be heard throughout the United States.*

minutes, I learned of a new RDS radio that Pioneer and Delco have developed for AM/FM reception. The new radio not only displays selected frequencies, but also reads out the related stations call letters and names of songs and their artists being played (obviously both transmitting and receiving stations must be RDS-equipped for this feature to work). Will RDS become popular? Only time will tell!

Additional stations on the AM band include: WWLS in Chicago (890 kHz), a pacesetter in rock and roll during past eras; WWGY of Schenectady, New York (810 kHz), a classic super station famous for talk shows and sports; and WSM, Nashville, Tennessee (650 kHz), famous for direct broadcasts from the Grand Ole Opry. Other stations of particular interest include: WWL, New Orleans (870 kHz), featuring talk shows and original jazz music; KCBS of San Francisco (740 kHz), the CBS affiliate; WABC, New York (770 kHz), ABC's affiliate; WRKO in Boston (680 kHz); and Radio Marti in the Florida Keys (1180 kHz). If you live in a remote area or travel by auto frequently, I encourage you to make a copy of Figure 2-3 and keep it by your radio for reference. The stations are good listening during normal times, and also informative "second and third" opinions during times of unrest.

Alternate News Sources

Talk shows and religious broadcast stations often provide additional information and insight to news covered on conventional AM/FM and TV facilities. Some talk shows are accurate and informative; others border on sheer gossip and sensationalism. There are no set references or regulations for what is broadcast by many of these programs, so *caveat emptor* and never depend on a single pair of talk shows for the "full picture."

Figure 2-4 *One of the most well known and respected religious broadcasters on television and shortwave is the Eternal Word Television Network, EWTN and shortwave station WEWN. The network is spearheaded by Mother Angelica (rear row, 10th from left/8th from right), and her group of devoted nuns.*

Religious-based programming, however, is a different matter. Although they consider subjects from religious viewpoints, they are usually very accurate in their details. Two of the most respected religious broadcasters are WWCR in Tennessee and WEWN in Birmingham, Alabama.

Generally stated, WEWN is a multiple media broadcaster utilizing television, shortwaves, and regular AM/FM media for the dissemination of information and the Catholic gospel.

EWTN's overall mission of spreading the eternal word and teaching people around the world about Christianity began with a satellite TV program around 1980. They expanded to include a worldwide shortwave broadcast station, WEWN, in 1993. Then during 1996, direct satellite feeds for AM and FM stations were included, allowing them to relay programming nationwide. As a result, many AM and FM stations nationally, indeed worldwide, carry segments between 30 minutes and several hours of EWTN daily or weekly. The broadcasts are in both English and Spanish, and include daily news with headline stories from a distinctly Catholic perspective in addition to devotions, prayers and a live daily Mass. The network's most popular program is "Mother Angelica Live," and it has an incredibly large following. The Eternal Word Television Network was founded by Mother Angelica of Our Lady of the Angels Convent in Birmingham, Alabama. It is a cloistered order of very devoted nuns and is one of the most powerful voices of the Catholic church today.

Alternative Talk Shows

Besides the run-of-the-mill talk shows on the radio networks, such as Larry King and Rush Limbaugh, there are others that carry conservative/religious and "the other side of the news" from their point of view. A few are found on the major networks, however, most have their own networks and transmit to their affiliates via satellite feeds and by the use of shortwave stations located in several areas of the U.S.

The satellite feeds from these talk shows can be monitored by some home satellite systems without the use of special receivers. On other services, a reasonably priced audio subcarrier receiver can be used for receiving these programs. These services are outlined in a subsequent chapter on Satellite Communications.

CHAPTER THREE

Shortwave Radio

A complete world of news directly from the source, information, and long-range communications of all types fill the shortwave spectrum of 1.7 to 30 MHz every day. Indeed, this "high frequency" range is literally a gold mine of ethnic programs and factual news from around the world—and it is readily available to survivalists in both metropolitan and remote areas alike. In addition to conveying a vast amount of international news, much of which may go unreported or heavily edited by the media in the U.S., shortwave is also "terra firma" for radio amateurs (hams), ships at sea, and aircraft on international flights. Shortwave also is used by numerous military, guerilla and clandestine forces, and many other services. Investing in a shortwave receiver and "SWLing" (shortwave listening) truly puts the world at your fingertips!

When tuning the shortwave bands during the evening hours, typically one might hear authoritative news from BBC London on 6175 kHz, alternate views from Croatian radio on 7370 kHz, and ham operators relaying information regarding a hurricane in the Gulf of Mexico to mariners on 14300 kHz. You also might hear ship-to-shore communications on 8784 or 8260 kHz, original New Orleans jazz music from WRNO on 9545 kHz (9.545 MHz), and aeronautical weather forecasts on 10051 kHz. Tuning further, you might uncover some mysterious underworld communiques such as "I am coming in at 12:15—no running lights—flash your headlights twice or I am out of there" on some oddball frequency between 6900 and 7000 kHz. Tuning a bit further, you might find El Salvador guerillas operating radio Venceremos around 6560 kHz, while anti-

Figure 3-1 *Shortwave receivers are available in a wide variety of styles and price ranges to fit every need and budget. Less expensive units lack SSB and CW mode reception, but are quite adequate for tuning in AM mode shortwave broadcast stations worldwide.*

Sandinistia Contras operate radio Quince de Scptimbre on 5950 kHz. If you are unfamiliar with the general concepts of SWLing, however, your first or second evening's tuning might result in only hearing Radio Moscow's strong signal on 7150 kHz or WWV's time signals on 5 or 10 MHz. Knowing what frequency to tune and at what time of the day or night is the key. Read on!

As you can logically surmise, the shortwave spectrum of 1.7 to 30 MHz is quite different from the regular AM and/or FM bands. Its vast range, for example, encompasses over a dozen international broadcast bands, numerous communications bands, and an unlimited number of channels for other services. Additionally, international shortwave broadcast stations shift bands and frequencies several times a day/night rather than staying on a single frequency as conventional AM/FM stations. Tuning in a specific station might seem like a hit-or-miss situation to the newcomer, but it is actually easy. The secret is to use an accurate program guide that lists stations, their transmitting frequencies and related times. These guides include the annual Worldradio Television Handbook (WRTH), monthly Monitoring Times magazine, and Popular Communications magazine (all available through shortwave stores and larger magazine and book dealers nationwide).

A mixture of transmission modes is used on shortwave. International broadcast stations typically use AM, ham radio operators, military services, airplanes and ships use single sideband (SSB). Other Amateur Radio activities are in Morse code (CW), and also there are many radioteletype (RTTY) and facsimile (fax) transmissions. Teletype transmissions are used by news services worldwide, and fax transmissions are used for conveying weather pictures. Receiving both RTTY and fax via your shortwave receiver is quite simple, especially if you have a home computer. Converter units, available from several companies nationwide, simply connect between your shortwave receiver's external speaker or "record" socket and the RS-232 port on your computer. Install the appropriate interfacing software in the computer, tune in the teletype and facsimile services frequencies, and you have your own in-home printout of direct news and weather.

Figure 3-2 *A shortwave monitoring post may consist of several receivers and tape recorders, or simply be a portable receiver used at any convenient time or location. Choice depends on individual needs and preference.*

Finally, the antennas used with shortwave receivers differ from those used with regular AM/FM receivers. For consistently good long-range reception of average to weak signals, a single outdoor wire antenna between 30 and 70 feet long does a respectable job. Alternately, and assuming you do not live in a metal enclosed building, a single long wire strung around a room in your house and connected to the external antenna socket or terminal on a shortwave receiver is a good choice. Some portable shortwave receivers have "pull up" antennas and they often prove adequate for receiving stronger stations' signals in the international shortwave broadcast bands. Additional information on each of the previously discussed topics will be included as we progress.

The International Shortwave Broadcast Bands

A multitude of independently owned, government backed, and church-affiliated stations in over 200 countries transmit programs to over 100 million listeners on shortwave every week. Indeed, every major country in the world in addition to many small countries beam their English language transmissions directly to North America. Their signals can be received with good clarity on even an inexpensive shortwave radio capable of AM mode reception. Some of the stations easiest to hear are the high-powered international broadcasters. Some examples of those super stations are BBC London, The Voice of America, Radio Moscow, HCJB Equador and Deutsche Welle Germany. Spotting these high-powered stations with just an old-style shortwave radio (those with a large analog dial and single tuning knob) is easy. You just zip across a band, and their signals rattle the speaker. Slightly below that upper level of signals are hundreds of other broadcast stations in various countries. Some examples of these stations are Radio Sweden, Vatican Radio, RNE Spain, and Radio Japan. Other popular shortwave stations include Radio Australia, Radio Thailand, the Voice of Vietnam, Radio Havana, WRNO New Orleans and WWCR in Tennessee. A wide variety of fascinating programming and music can be heard from all of the previously mentioned stations throughout the Americas. Several shortwave stations use relay bases for coverage into their targeted areas during prime listening hours. The BBC, for example, is often relayed via Radio Canada, and The Voice of America has independent relay bases on all continents. Radio France and Radio Nederland also utilize relay sites in former colonies in the Caribbean (and elsewhere) on an exchange basis. All of the previous broadcasts are carried within the international shortwave broadcast band: each station operating on specific frequencies established by the World Administrative Radio Conference (WARC). These conferences are held every few years in Geneva, Switzerland and determine what frequencies are used for what services. Now let's briefly discuss each of the popular international broadcast bands (see Figure 3-4).

Figure 3-3 *One of the most well-known medium power international shortwave stations heard consistently in the United States is Radio Japan. Reports of reception from listeners are often rewarded with impressive "QSL" cards.*

Frequency in Kilohertz (kHz)	Frequency in Megahertz (MHz)	Meter Band
2300 - 2500	2.3 - 2.5	120
3200 - 3400	3.2 - 3.4	90
4600 - 5100	4.6 - 5.1	60
5950 - 6200	5.95 - 6.20	49
7100 - 7300	7.1 - 7.3	41
9500 - 9900	9.5 - 9.9	31
11650 - 12050	11.65 - 12.05	25
13600 - 13800	13.6 - 13.8	22
15100 - 15600	15.1 - 15.6	19
17550 - 17900	17.5 - 17.9	16
21450 - 21850	21.45 - 21.85	13
25670 - 26100	25.67 - 26.10	11

Figure 3-4 *Popular international shortwave bands, arranged by frequency and meters. Stations of all nationalities daily beam their English transmissions to the United States on these bands.*

2.300 - 2.500 MHz (120-meter band)

This lowest frequency range is often called the tropical band, and it is used primarily by low-power stations for local broadcasting. Reception of Latin American stations and maybe some South Pacific stations can be quite challenging yet equally rewarding. Quality will vary from day to day and season to season. This is also a late night band "open" between 9 p.m. and 5 a.m. your local time. An outdoor antenna approximately 100 feet long or a loop antenna especially tuned for low frequency use is required for long-distance reception.

3.200 - 3.400 MHz (90-meter band)

This "middle" tropical band is also used by low-power stations for local broadcasting, however it supports more stations than the 120-meter band. The listening, again, reflects local culture and interests, and stations are usually located in Latin America or the South Pacific. Reception is best during late night hours and a good outdoor antenna is beneficial.

4.600 - 5.100 MHz (60-meter band)

An upper tropical band that also supports many low-power and local area shortwave stations in Latin America in addition to Pacific and Asian areas. Assuming late night tuning using a good receiver and antenna, stations in several countries can be received reliably year round. Static levels, however, are more predominant on all three tropical bands during summer. Also noteworthy, many tropical band stations reward reports of reception from distant listeners with elaborate QSL cards, pens and pendants. The original romance of shortwave listening is truly alive and well and thriving on the tropical bands worldwide.

5.900 - 6.250 MHz (49-meter band)

This is the most popular shortwave band used for international broadcasting, and it supports a large number of stations in countries on all continents. This band typically "opens" around dusk and stays open until

Figure 3-5 *A very popular and easy to receive tropical band shortwave broadcast station is La Voz Evangelica, HRVC, in Honduras.*

around dawn. An outdoor antenna between 40 and 60 feet long provides good reception of weaker signals. Portable radios with simple pull-up antennas are capable of receiving stronger stations on the 49-meter band.

7.100 - 7.500 MHz (41-meter band)

This band also is used for international broadcasting and is shared with amateur radio services. Basically a nighttime band like 49 meters, except "opens" an hour or so earlier in the afternoon and "closes" an hour or so later after sunrise. This band is not used for broadcasting in the United States, so stations heard here are located in foreign countries.

9.400 - 10.000 MHz (31-meter band)

The 31-meter band is another popular international band used by broadcasters worldwide during both day and night. The best listening time is between 3 p.m. and 10 a.m. your local time. It is ideal for tuning in broadcasts during prime-time evening hours. It is often crowded with numerous stations.

11.600 - 12.100 MHz (25-meter band)

The 25-meter band is probably the best all-around, all-time, all-season band. It is crowded with stations of all nations. The reception is good during the day and continues well into the night during the summer months. A 20 to 40 foot long-wire antenna is good for receiving weaker stations on this band (also works well on the 31-meter and 22-meter bands).

13.600 - 13.900 MHz (22-meter band)

The 22-meter band was originally assigned for international broadcasting by WARC a few years ago; however, many other stations are now appearing in this upper-frequency range. Activity will surely increase during the coming years. 22 meters is primarily a daytime band. Considering its close proximity to amateur radio's famous long-range band of 20 meters, the 22-meter band is extremely promising for future broadcasting.

15.100 - 15.600 MHz (19-meter band)

This is another very popular band used by international broadcast stations worldwide. Reception is good during daylight hours and also extends into evening hours during the summer. An impressive band for receiving long-range signals.

17.600 - 17.900 MHz (16-meter band)

The 16-meter band is another long-distance daytime band that is becoming increasingly popular among shortwave stations and listeners worldwide. This band, and the following 13- and 11-meter bands, are especially good during years of high sunspot activity. They are also appealing because shorter antennas provide optimum reception. Typically an antenna for 16, 13 or 11 meters will be only 8 feet long.

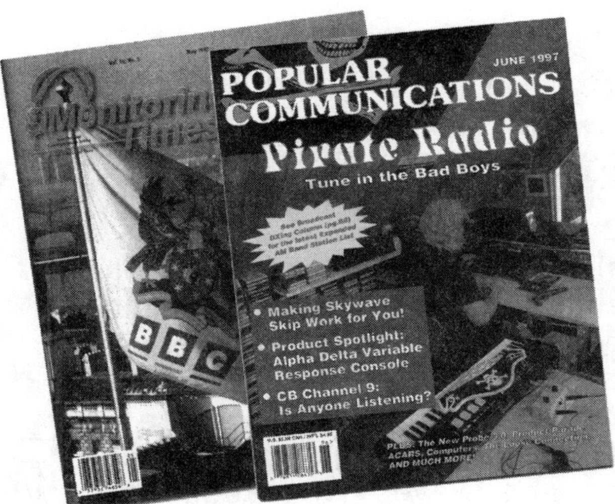

Figure 3-6 *Two popular magazines with up-to-date information / frequencies regarding shortwave broadcasters is Popular Communications and Monitoring Times.*

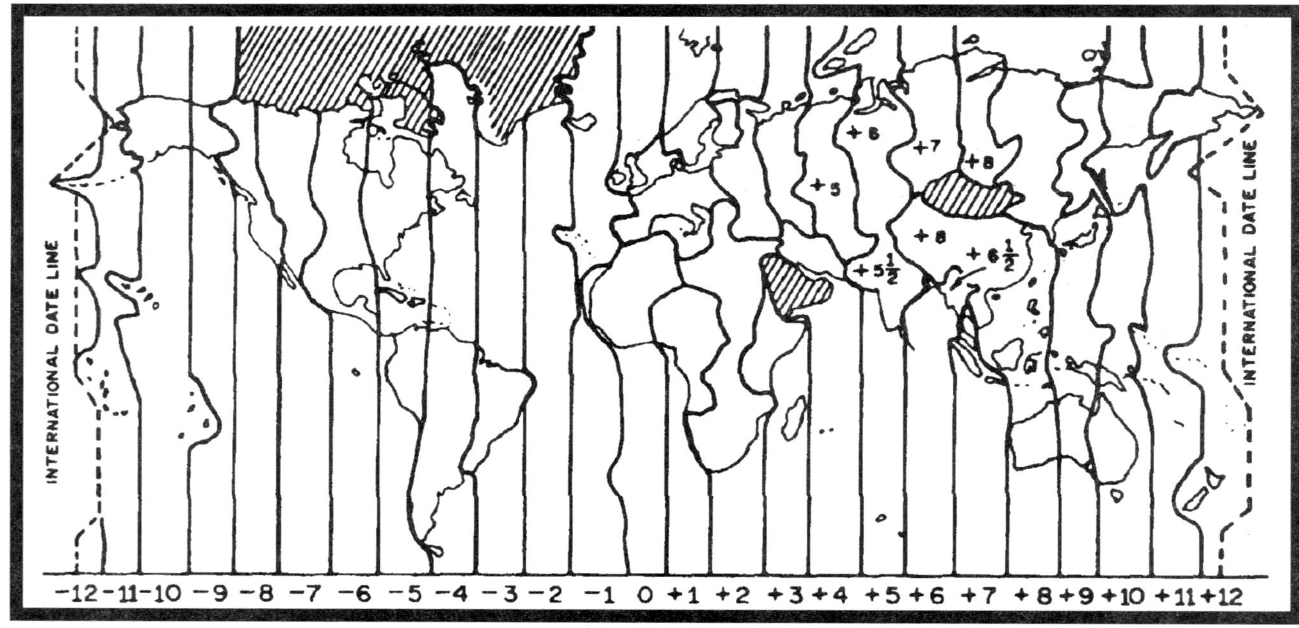

Figure 3-7 *Universal Time (UTC) areas worldwide, often called Greenwich Mean Time (GMT). These times will be the same the world over. For example, in any time zone the GMT and UTC time will be the same.*

21.450 - 21.850 MHz (13-meter band)

This is a daytime-only band that sees little use during years of low sunspot activity and higher use with very good long-range reception during years of high sunspot activity.

25.600 - 26.100 MHz (11-meter band)

An upper HF daytime-only band for international broadcast stations. During years of high sunspot activity, signals can be received with surprisingly good signal strength between 9 a.m. and 3 p.m. local time.

Before continuing, let's pause to explain the meaning of meter bands on shortwave. The frequency (kilohertz or megahertz) on which a radio station transmits refers to the number of complete waves it sends

UNIVERSAL TIME (UTC) OR (GMT)
LOCAL TIME ZONES, USA

UNIVERSAL TIME (UTC)	EDT	EST/CDT	CST/MDT	MST/PDT	PST
0000	8 p.m.	7 p.m.	6 p.m.	5 p.m.	4 p.m.
0100	9 p.m.	8 p.m.	7 p.m.	6 p.m.	5 p.m.
0200	10 p.m.	9 p.m.	8 p.m.	7 p.m.	6 p-m.
0300	11 p.m.	10 p.m.	9 p.m.	8 p.m.	7 p.m.
0400	Midnight	11 p.m.	10 p.m.	9 p.m.	8 p.m.
0500	1 a.m.	Midnight	11 p.m.	10 p.m.	9 p.m.
0600	2 a.m.	1 a.m.	Midnight	11 p.m.	10 p.m.
0700	3 a.m.	2 a.m.	1 a.m.	Midnight	11 p.m.
0800	4 a.m.	3 a.m.	2 a.m.	1 a.m.	Midnight
0900	5 a.m.	4 a.m.	3 a.m.	2 a.m.	1 a.m.
1000	6 a.m.	5 a.m.	4 a.m.	3 a.m.	2 a.m.
1100	7 a.m.	6 a.m.	5 a.m.	4 a.m.	3 a.m.
1200	8 a.m.	7 a.m.	6 a.m.	5 a.m.	4 a.m.
1300	9 a.m.	8 a.m.	7 a.m.	6 a.m.	5 a.m.
1400	10 a.m.	9 a.m.	8 a.m.	7 a.m.	6 a.m.
1500	11 a.m.	10 a.m.	9 a.m.	8 a.m.	7 a.m.
1600	Noon	11 a.m.	10 a.m.	9 a.m.	8 a.m.
1700	1 p.m.	Noon	11 a.m.	10 a.m.	9 a.m.
1800	2 p.m.	1 p.m.	Noon	11 a.m.	10 a.m.
1900	3 p.m.	2 p.m.	1 p.m.	Noon	11 a.m.
2000	4 p.m.	3 p.m.	2 p.m.	1 p.m.	Noon
2100	5 p.m.	4 p.m.	3 p.m.	2 p.m.	1 p.m.
2200	6 p.m.	5 p.m.	4 p.m.	3 p.m.	2 p.m.
2300	7 p.m.	6 p.m.	5 p.m.	4 p.m.	3 p.m.

Figure 3-8 *With this chart you can convert your local time to UTC—UTC is 4 hours ahead (+) of Eastern Daylight Time; 5 hours (+) ahead of Eastern Standard Time and Central Daylight Time; 6 hours ahead (+) of Central Standard Time and Mountain Daylight Time; 7 hours ahead (+) of Mountain Standard Time and Pacific Daylight Time; and 8 hours ahead (+) of Pacific Standard Time. Add one hour to local time when Daylight Savings Time is in effect.*

out each second. The physical length of each wave is measured in meters, and is inversely proportional to the frequency. As frequency increases, wavelength decreases (and vice versa). Generally speaking, frequency can be converted to wavelength using the formula: meters = 300,000 divided by kilohertz. As an example, 300,000 divided by 49 = 6122.44 kHz. Likewise, frequencies can be converted back to meters by dividing them into 300,000. Example: 300,000 divided by 6122.44 = 49 meters. If you spend a few minutes inserting various frequencies into the formula, you will discover each one corresponds to a fractionally different wavelength or "meter band." For simplicity, however, they are generally rounded off and described as "the 25-meter band or the 31-meter band or the 49-meter band." In other words, "meter band" gives an approximate frequency while kilohertz or megahertz gives an exact frequency.

American Shortwave Listener's Club (ASWLC), 16182 Ballad Lane, Huntington Beach, CA 92649. Broadcast, utilities. Monthly bulletin. $18 per year.

Association of Clandestine Enthusiasts (ACE), P.O. Box 46119, Baton Rouge, LA 70895. Pirates, clandestines, numbers stations. Monthly bulletin. $12 per year.

Association of DX Reporters, 7008 Plymouth Rd., Baltimore, MD 21208. Broadcast, utilities, ham bands. Monthly bulletin. $15 per year.

Australian Radio DX Club, P.O. Box 227, Box Hill Victoria 3128, Australia. Broadcast and utility. Monthly bulletin. $A-48 per year.

Canadian International DX Club, #61-52152 Range Road 210, Sherwood Park, Alberta T8G 1A5. Broadcast, utilities, ham bands. Monthly bulletin. $18 per year.

Danish Shortwave Club International, Tavlaeger 31, DK-2670 Greve Strand, Denmark. Broadcast, utilities. Monthly bulletin. $23 per year.

DX Australia, P.O. Box 285, Mt. Waverly, Victoria, 3149. Broadcast. Monthly bulletin. $30 per year.

New Zealand Radio DX League, P.O. Box 1313, Invercargill, New Zealand. Broadcast. Monthly bulletin. $20 per year.

Figure 3-9 *Clubs for shortwave listeners are ideal for staying up to date on aired activities, hot frequencies, band conditions and new radios. Clubs listed here offer introductory information plus newsletter sample for $2.00 to cover copying and mailing costs.*

International shortwave broadcast stations move or shift frequency every few hours to take advantage of signal skip/propagation into their targeted area. A station may use a frequency in the 19-meter band during the day, shift to 25 meters around dusk, and then move down to 49 meters during the late evening. The next day, that station would again "migrate" upward in frequency. Selecting optimum frequencies is a science, and some stations also move within a particular meter band on an hourly basis and according to seasons of the year. Additionally, some of the lower powered pirate stations often shift frequency or appear on unexpected frequencies in a random manner to sidestep jammers. This technique is particularly popular when the broadcasts include propaganda or counterpropaganda. Up-to-date frequency guides regarding "who is where and when" are published in monthly magazines such as Popular Communications and Monitoring Times.

As stated earlier, do not expect shortwave stations to "stay put" on a single frequency hour after hour, day after day, month after month. Also, remember most frequency guides for shortwave list stations according to Greenwich Mean Time (GMT) or Universal Time (UTC). Converting this 24 hour time to your local time is relatively easy: the hours of 0100 to 1200 correspond to 1 a.m. until 12 noon. The hours of 1300 to 2359 correspond to our "past 12 noon," or 1 p.m. until 11:59 p.m. Midnight is 0000. The

Greenwich Meridian passes through Europe; therefore, Greenwich Mean Time or Universal Time is 5 hours ahead of Eastern Standard Time. In other words, 2300 GMT equals (subtract 5 hours) 1800 EST, or 6 p.m. EST. For the Central time zone, 2300 GMT is (subtract 6 hours) 5 p.m. CST. The only complexity results when Greenwich Mean Time goes beyond an evening and into the next day. For example: 0200 GMT Tuesday equals (again subtract 6 hours) 2000 or 8 p.m. CST on Monday evening. This is simply because Tuesday started at 0001 GMT.

Finally, we recommend serious SWL enthusiasts expand and share their knowledge by joining one or more of the popular SWL clubs (see Figure 3-9). Comparing notes with others always beats listening alone.

The Two-Way Communications Bands

Many two-way communications bands are liberally sprinkled between the international broadcast bands and throughout the shortwave spectrum. Some of these bands are sectored into "channels" similar to the concept used on Citizens Band. Others are a range of frequencies used in a more "tune and call" manner. Marine, aircraft, and Amateur Radio services use the two-way bands, but unlike the international broadcasters, do not follow specific time and frequency schedules. These lower power services also have slightly weaker signals than the international broadcasters. This makes them more of a challenge to copy (i.e., receive). They use different and more efficient transmission modes such as single sideband (SSB) and Morse code (CW). A good communications-grade shortwave receiver with an outdoor antenna (rather than a portable shortwave receiver with a pull-up antenna) is suggested for two-way communications reception. Such an investment is well worth the cost, as a sheer wealth of information for survivalists of all interests is conveyed within these ranges. Scattered on "odd ball" frequencies between and around both two-way and international broadcast bands are numerous other shortwave activities. These include guerilla and antigovernment communications, clandestine activities, smugglers, illegal fishing communications, and pirate broadcasters of various convictions. Finally, a number of often overlooked frequencies throughout the shortwave range are used for radio teletype transmissions of news and facsimile communications of pictures and weather charts. Shortwave truly supports a variety of activities capable of holding one's interest for two lifetimes!

The Marine Bands

Mariners traveling more than 20 miles off shore and ships on the high seas rely on SSB in specific HF bands for communicating with other ships and shore-based stations. Overall, there are more than 250 international and domestic channels established for long-range marine communications. Some of the channels are "simplex," which means both ship and shore stations transmit and receive on the same frequency. Other channels are "duplex," which means separate transmission frequencies are used for ships and coast stations. Listening to "both sides" of such conversations requires a little know-how. Assuming your receiver has dual VFOs, set one to each frequency and quickly switch between VFOs to hear both parties. Alternately, frequency pairs for each channel can be loaded into adjacent memories which you can switch between in a rapid-fire manner to hear both sides of conversations. Some of the newer and more elaborate communications receivers include "dual watch reception," which means they are similar to two receivers in one cabinet. If you own one of these units, set it to receive both frequencies and use its balance control to set both volumes at the same level. This is top-notch monitoring indeed! Communications transceivers used aboard ship and by shore operators are also set up according to channels (although some transceivers display both channel and frequency), therefore, operators often refer to their frequency as "channel XXX." A few channels within each band are also set aside for ship-to-shore telephoning. In this case, an initial call brings a response from a shore operator who records charge card numbers, etc. and places the telephone call for the distant party. Channels used for marine telephoning are also subject to change from time to time.

When tuning marine bands or channels, bear in mind that activity is on a strictly as-needed or eclectic basis. Do not expect to hear in-progress communications nonstop! A clever idea is to load a few frequencies/channels from each marine band into your receiver's memories, then scan the memories until activity is spotted. Remember that frequencies and channels below approximately 10 MHz are used more for nighttime communications, while frequencies/channels above 10 MHz are primarily used for daytime communications. Before leaving our discussion of marine bands, here are three Coast Guard channels you may find interesting to monitor (1) Receive: 4.428 MHz, Transmit: 4.134 MHz; (2) Receive: 8.765 MHz, Transmit: 8.241 MHz; (3) Receive: 13.113 MHz, Transmit: 12.342 MHz. Coast Guard simplex frequencies used for emergencies and search/rescue operations are often conducted on the simplex channels of 7,375 kHz, 9074 kHz and/or 11028 kHz.

The Aircraft Bands

Both private and commercial aircraft on flights within the United States typically use VHF channels for communicating up to 250 miles. Line of sight is quite good when your antenna is up 20,000 feet! When traveling off shore, internationally, or "out of VHF/line of sight range," however, frequencies and channels within the HF shortwave spectrum are employed. Some of these communications are on set frequencies, others are selected by the parties using them (possibly for military and security purposes). A general list of their bands and ranges of operation are shown in Figure 3-10. Remember to tune frequencies below approximately 10 MHz during evening hours and frequencies above 10 MHz during daytime hours. Lucky aircraft monitors may be fortunate enough to hear military tactical operations in foreign lands as they take place. Good luck and good tuning for them!

Aeronautical Stations		
2.300 - 3.500	9.950 - 11.650	17.360 - 17.750
4.000 - 5.950	12.050 - 12.230	17.900 - 21.000
6.200 - 7.000	13.200 - 13.600	21.850 - 22.000
8.815 - 9.500	14.350 - 15.100	22.855 - 25.010

Figure 3-10 *HF range aeronautical channels used by aircraft on international flights.*

Other Communications

In between, and often near international broadcast bands and two-way communications bands, are the various seldom-tuned frequencies used by guerillas, clandestine forces, pirate radio broadcasters and various utility services. Some of these use SSB or "cloaking," others use AM in hopes of capturing an audience for their propaganda. Tuning in the stations is probably the greatest cat-and-mouse game in shortwave radio, because the selection of frequencies can be unpredictable. Guerillas and clandestine operations have been known to use the range of 6900 to 7000 kHz and are also found above the high end of the 25-meter international broadcast band. Pirate radio broadcasters are typically low-power AM stations using older modified amateur radio transmitters. Many of the larger pirate stations use a lavish array of home-built equipment and have been found aboard sea vessels no longer in commercial service and "parked" slightly offshore. Some are located in isolated areas of the U.S. and other countries. Utility services include various fixed or land-based stations and two-way communications link in the HF spectrum, and basically do not conform to a specific band plan. Additional "how to find them" tuning guidance is found in monthly publications like Monitoring Times magazine and Popular Communications magazine—both available through larger magazine stores and amateur radio/shortwave radio equipment dealers nationwide.

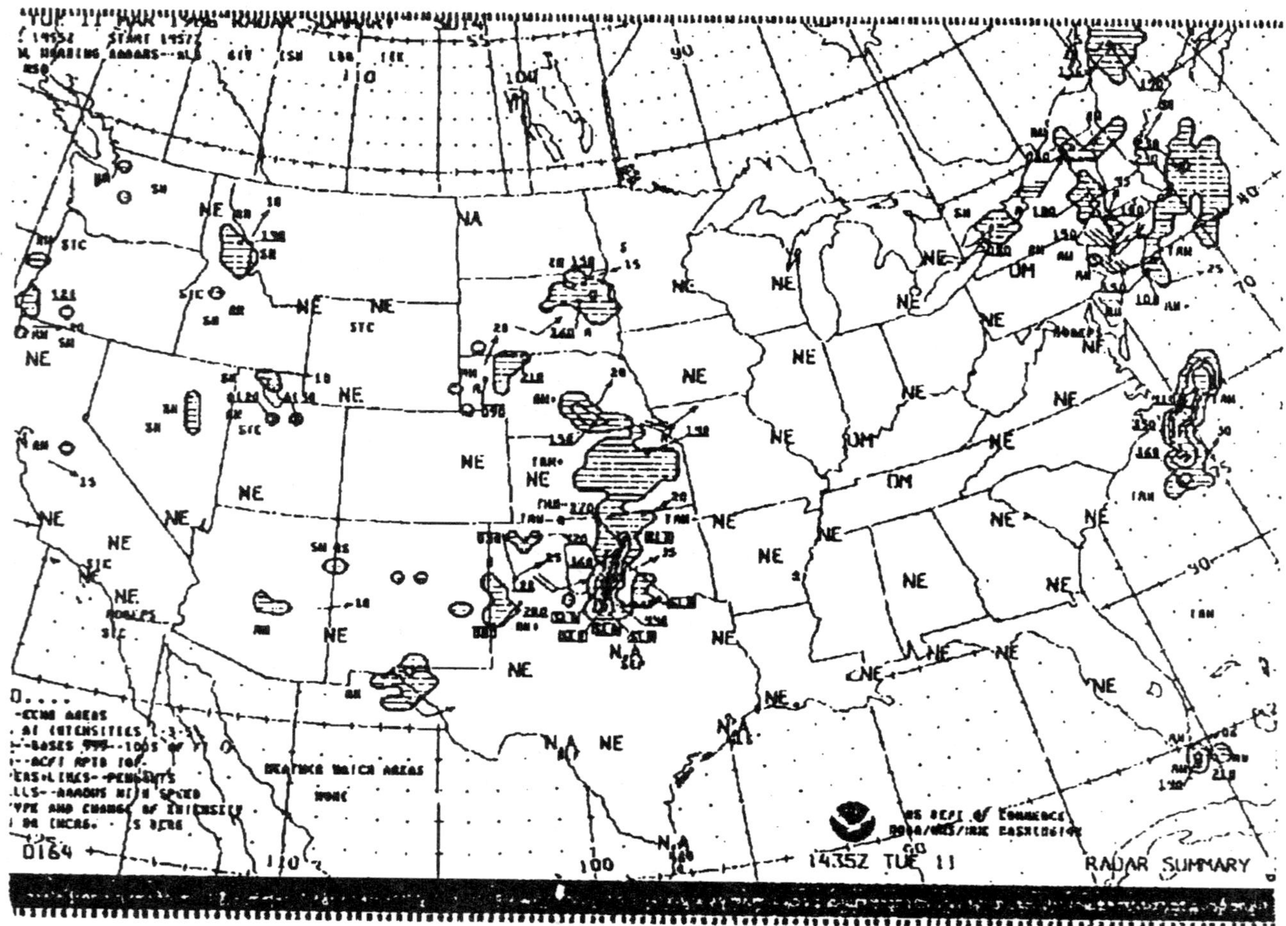

Figure 3-11 Sample weather fax picture received off HF bands.

Among the more interesting utility communications on shortwave are radioteletype and facsimile transmissions. Teletype is often used for news agencies and is a complete story in itself. Several RTTY tuning books listing current frequencies are available from amateur radio equipment dealers nationwide. Likewise, weather fax transmissions are easy to receive and can be printed right in your home. An example of a fax weather picture received via HF shortwave is shown in Figure 3-11. A few popular frequencies for weather fax reception are: 9157.0 kHz/WLO Mobile, Alabama; 9203 kHz/GFE Bracknell, England; and 5093 kHz/ LZJ Sofia, Bulgaria. Military fax transmissions are often noted around 8080 kHz/NAM Norfolk, Virginia; 7530 kHz/NMF Boston, Massachusetts; and 21839 kHz/NTM Pearl Harbor, Hawaii.

Reception of RTTY or fax requires a high-grade communication receiver and an add-on computer interface or dedicated converter unit for displaying pictures and text. Computer interface systems are popular and low cost (assuming you already have a home computer); however, dedicated units provide noticeably superior copy of weak and strong signals alike. RTTY/fax converters and interface units are available from shortwave and amateur radio equipment dealers nationwide. Can shortwave communications receivers also be used to tune the Citizens Band? Yes, provided they cover the range of 26 to 28 MHz and have good sensitivity on those higher frequencies. A kilohertz-to-CB channel number conversion chart is shown in a following chapter regarding CB. Is there a disadvantage? Sure, you cannot talk back if you only have a receiver!

Hams operate within the following ranges:

1.800	-	2.000 MHz	........160 meter band
3.500	-	4.000 MHZ	.... 80 meter band
7.000	-	7.300 MHz	.... 40 meter band
10.100	-	10.150 MHz	.... 30 meter band
14.000	-	14.350 MHz	.... 20 meter band
18.068	-	18.186 MHz	.... 17 meter band
21.000	-	21.450 MHz	.... 15 meter band
24.890	-	24.990 MHz	.... 12 meter band
28.000	-	29.700 MHz	.... 10 meter band

Figure 3-12 *Popular Amateur Radio HF bands listed according to frequencies and meters.*

The Amateur Radio Bands

Surely the most informative, unique, and oftentimes curious communications on shortwave are conducted among amateur radio operators around the world. Indeed, Amateur Radio's HF bands (shown in Figure 3-12) are often the heartbeat of action and news as it happens during both normal daily routines and extraordinary occurrences of all kinds. Tuning the amateur radio bands (again those below 10 MHz during evening hours and those above 10 MHz during the day) one might hear casual conversations between retirees in Florida and friends in New England, British or Far East amateurs chatting with other amateurs throughout the United States, and "aeronautical mobile" on a military flight, and more. Tuning to 14.300 MHz, you may hear ham operators on small boats talking with other hams ashore. Tuning 14.282 MHz, you might hear hams relaying current information regarding a hurricane approaching the coast. Tuning the "DX range" of 14.160 MHz to 14.225 MHz, you may hear "pileups" of enthusiastic amateurs rapidly contacting rarely received operators in Kuwait, Bangladesh, and almost any other areas of the world. Tuning the lower 100-kHz range of any Amateur Radio HF band, you will hear a fascinating array of Morse code

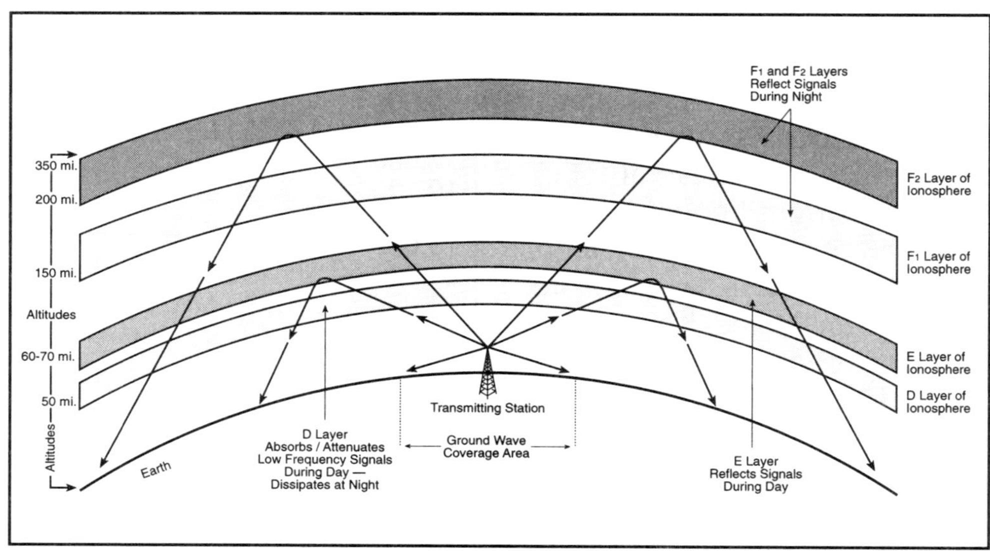

Figure 3-13 *Various layers of the earth's ionosphere influence long range reception on HF bands.*

(CW) communications. Connect a cable between one of the previously mentioned RTTY/fax/CW converter or interface units and your radio's speaker output and computer or printer input, and you can read Morse code transmissions directly in real time. The world of Amateur Radio is far more extensive than we can describe here, so additional information is presented in a following chapter. Consider this an open invitation to read more about Amateur Radio, visualize its endless applications to your own life, get an F.C.C. license and join us on the air! Need more encouragement? Many amateur radio transceivers include full coverage reception of shortwave on all modes, and are equal to some of the best stand-alone shortwave receivers. "Getting everything in a single cabinet" is a deal that is difficult to turn down!

How Shortwave Signals "Skip" Worldwide

How are shortwave radio stations able to reach out around the world while regular FM broadcast stations can only cover a 100- to 150-mile range? Stated in a single word, the answer is propagation. Signals on frequencies between 500 kHz and 30 MHz consist of both ground waves and skywaves. The ground waves travel only a short, line-of-sight distance; however, the skywaves are reflected by gaseous layers above the earth and bounce back into distant locations. Ground waves radiated by FM band signals reach out well into local areas; however, their skywaves go straight through the earth's gaseous layers and continue into space where they are lost. The layer of gas above the earth responsible for reflecting radio signals is called the ionosphere. It receives ultraviolet energy from the sun. This energy causes the ionosphere to act as a radio mirror that reflects signals over great distances. When the signal returns to earth at a distant point, it may reflect off water or land and bounce back to the ionosphere where it is once again reflected back to earth at an even greater distance from its point of origination. After two or three "bounces," the signal has literally traveled around the world.

Technically speaking, the ionosphere is comprised of four layers or densities (see

Figure 3-13). These layers are designated D, E, and F1 and F2 layers. The D layer is closest to the earth, approximately 50 miles up. In reality it does not bounce or reflect signals but rather absorbs lower frequency signals. The D layer is present during the day but rapidly disappears during the night when ultraviolet energy from the sun is not present. When the D layer has dissipated, upper layers are available for reflecting radio signals at night. The next layer up is the E layer, which is located approximately 60 miles above the earth. This is the main layer that reflects most shortwave stations' signals back to earth during the day. When the D layer becomes weaker at night, it permits signals below 12 MHz to reflect off the higher F1 and F2 layers. The F1 and F2 layers are typically between 100 and 200 miles above the earth and are responsible for reflecting ultra-long-distance signals at night. During that time, the two F layers actually combine into a single layer.

Exact ionospheric layers and precise frequencies are not used in our discussion. That is because the ionosphere's layers are constantly changing according to ultraviolet energy from the sun that varies day to day and month to month. It also varies within an 11-year solar cycle. Describing all the variables associated with the ionosphere and its reflection of signals is a complete study in itself, so let's simplify the matter with some easy to understand facts for SWLing. First, understand there will always be good and bad days for receiving signals over long distances. Unexpectedly, signals from distant stations randomly fade but eventually return with even greater signal strength. As a general tuning guide, consider nighttime hours best for receiving signals from long distances via the shortwave bands between 1.7 and approximately 9 to 12 MHz. Likewise, frequencies between 12 MHz and 30 MHz reflect off the ionosphere best during daylight hours. Do you realize daytime hours are also good for receiving over long distances? Yes, indeed they are. We have been conditioned to assume that night hours were the best time for long-distance reception because of our familiarity with the AM broadcast band.

Summarizing the previous statements, the lower shortwave frequencies provide the best reception during night hours, and the higher frequencies provide the best reception during daytime hours. Also, remember the ionosphere is continually heated and cooled by the sun's energy, thus its ability to reflect radio signals changes almost continuously.

Think back a page or two and you will now understand why international shortwave broadcast stations change frequencies every few hours—they are taking advantage of skip conditions in the ionosphere to reach their targeted area. "Riding the waves of ether" and bouncing signals off our earth's invisible mirror may seem like a game of chance, but it is actually a well-planned science.

Shortwave Receivers

Radios covering the shortwave bands are available in a wide variety of styles, sizes and price ranges to fit everyone's preference. Some models require 120-volt AC (home) power for operation, some use batteries, and some include 3-way power selection (home, auto and battery). One model even sports a hand crank you wind for 20 to 30 seconds and the built-in generator then powers the receiver for approximately 30 minutes (see our chapter on self-powered radios). The overall selection of shortwave radios is vast (newcomers might say "confusing"), and we can honestly say there are at least one or two models available to fit anyone's lifestyle.

Looking closer, we also find there is a somewhat vague dividing line between general consumer shortwave radios (usually portables) and communications-grade shortwave receivers (larger desktop units). This dividing line may be roughly defined according to price and modes of reception. Units priced below $500 are usually consumer-grade radios. Many of them feature only AM and FM modes of reception. Shortwave receivers priced above $500 usually include AM and FM modes of reception, a beat frequency oscillator (BFO) for SSB and CW reception, and more internal circuitry for copying (receiving) weak signals. Communications-grade receivers usually cost upwards of $1,000, but two encouraging notes should be mentioned at this point. First, most serious SWLers use receivers in the $500 to $1000 category and receive even weak stations quite successfully. Second, used communications receivers are often found at low prices at amateur radio dealers and ham radio conventions (hamfests) nationwide.

If you are only interested in receiving stronger international broadcast stations and casually tuning for pirate broadcasters, a portable shortwave receiver is good. If you are more interested in receiving any and all types of shortwave stations and signals, a high-performance communications receiver is definitely required. Regardless of which kind of shortwave receiver you select, I cannot overemphasize the importance of initially reading its manual. No one is more familiar with a particular radio than the people who manufactured it and wrote the manual. Read the manual at least twice, make certain you understand the explanations, and follow the directions given in the manual. It is an ideal guide for familiarizing yourself with a particular radio.

Let's discuss one of the more important aspects and features to consider when selecting a shortwave radio (portable or communications type). First, check to ensure that its coverage includes the range of frequencies you desire. Some portables cover only international broadcast bands, for example, while other portables squeeze several bands into a small tuning area making single-knob tuning quite tedious. Second, check to ensure that the unit is capable of receiving the signal modes you desire to receive. Third, check its source of power and optional power arrangements. If you are depending on a radio when AC power is not available and the unit will not operate on batteries, it is of little benefit.

Next, consider the technical aspects of the unit, such as frequency coverage, sensitivity, selectivity, band spread and type of frequency display. Full coverage from 500 kHz to 30 MHz is desirable, provided all-mode reception and good sensitivity on more critical upper frequencies (those above 14 MHz) are also present. Personally, I shy away from receivers that include FM band (88 to 108 MHz) coverage as they seem more "consumer grade" oriented. My reasoning is that different circuitry is required for both VHF reception and HF circuitry, and therefore, overall performance must be compromised.

Sensitivity refers to a receiver's signal-to-noise ratio, or ability to receive weaker signals. Generally stated, receivers with less than 0.5 microvolt sensitivity (that means one microvolt or two microvolts for 10 dB S/N ratio) are in the "consumer grade" category and receivers with better than .5 microvolt sensitivity (in other words .25 microvolt or .1 microvolt) are in the "communications grade" category.

Selectivity refers to the receiver's ability to separate stations on adjacent frequencies. Selectivity is usually measured in kilohertz at the -6 dB and -60 dB points on the receiver's IF response curve. If the ratio of those two bandwidths are greater than 2:1, the unit will be susceptible to interference from strong stations adjacent to your selected frequency. Communications-grade receivers usually include front panel adjustments for varying selectivity or rejecting adjacent channel interference. Sometimes such controls are labeled bandwidth, many times they are labeled IF shift, and other times they are labeled passband tuning. Some of the better receivers also include an IF Notch filter that can be adjusted across the receiving range to "null out" heterodynes or howls.

Simply explained, bandspread refers to how many kilohertz of coverage are obtained in one revolution of the main tuning knob. The greater the bandspread, the lower the kilohertz per dial revolution, and the easier it is to tune in stations. Lack of good bandspread requires a steady tuning hand. Further, warm-up drift and physical movement may influence the reception frequency. Strive to acquire the greatest bandspread possible in the price range receiver you can afford. Finally, frequency readout is important for tuning in exact stations with minimum effort. Receivers with conventional analog dials are acceptable for general tuning and casual listening; however, precise digital frequency readouts are much more desirable for zipping directly to an exact frequency (assuming the receiver is expensive enough to be accurately calibrated and take advantage of the digital display).

As you look closer at the front panel of shortwave receivers, some of their controls may seem confusing. Let's briefly review some of the more common knobs and push buttons.

The **Power On/Off** button turns the radio on or off. This control is sometimes separate and at other times included with the volume control.

Volume or AF (Audio Frequency) Gain increases or decreases the audio volume from the speaker.

RF (Radio Frequency) Gain essentially controls the overall sensitivity of the receiver. Clockwise settings permit receiving weaker signals; counterclockwise usually limits sensitivity to weaker signals.

Main Tuning selects the frequencies that the radio receives. Frequencies are usually shown on an associated dial or readout.

Up/Down buttons are used on some radios to shift frequencies higher or lower or to change bands.

RIT (Receiver Incremental Tuning) allows the exact frequency to be varied plus or minus one or two kilohertz without disturbing the main tuning knob setting.

Mode Switch selects AM, SSB, CW, RTTY, and other modes. Sometimes this switch is included as a series of push buttons or a single push button with sequential selection of modes.

Band selects a group of frequencies or "band" through that the radio will tune. May be equal to a range such as 5 to 10 MHz, a single megahertz spread, or a particular band (i.e., 49 or 31 meters). On older receivers, this control may be labeled "A, B, C" with each letter corresponding to a rather wide range of frequencies.

Selectivity or Bandwidth determines how wide or narrow the receiver's single-signal reception range is at a selected dial frequency. It may be labeled as 12, 6, 2.4, or 0.5 MHz. Wider settings provide the best quality audio for AM reception. Mid-range settings are good for single sideband and rejecting adjacent channel interference. More narrow settings are used for CW reception.

Passband Tuning or IF Shift permits varying a center frequency or a specifically-selected frequency. It also influences the bandwidth of the receiver. Overall, passband tuning provides good rejection of interference.

Attenuator reduces the signal strength of exceptionally strong signals to avoid overloading the receiver's front end "circuitry."

Preselector is a panel-selectable RF amplifier stage that increases the sensitivity of the radio.

Noise Blanker is a circuit designed to reduce noise produced by electrical interference such as leaky electric power lines, automobile ignition noise, and electric motor brush noise.

Automatic Gain Control (AGC) is an internal circuit that automatically controls the RF gain or overall sensitivity of the receiver. It is quite useful during fades. AGC increases receiver gain when the signal strength drops, and it decreases receiver sensitivity when the signal strength increases.

Memory provides an electronic storage medium for a particular frequency. It also provides instant recall of a desired frequency without manual tuning. Simple memories store a frequency and maybe a mode (each), but require dial selection of new frequencies (in other words the memory is not tunable). More elaborate memories recall their stored frequency, then can be retuned to any new frequency, and the frequency stored by pressing a single button.

Scan initiates automatic tuning through a pre-selected range of frequencies or a group of memories. Frequencies are sampled and displayed on the readout in rapid succession. Scanning speed is usually adjustable. Scanning can usually be set to stop on the reception of a signal or continue nonstop across the programmed receiving range or bank of memories.

Earphone Socket provides a front panel receptacle for earphones for private listening. Plugging in earphones usually disconnects a receiver's internal speaker.

Record Socket provides a connection for an external tape recorder. This connection usually maintains a constant volume level that is not affected by the receiver's AF Gain control. It allows you to record off-the-air transmissions while listening on the radio's internal or external speaker.

External Antenna Socket is usually a rear-panel socket designed to accept a PL-258 plug. Some receivers use screw terminals instead of a coaxial socket. Note: coaxial sockets use the shield (cable's outer section) as ground; center terminal/receptacle is "hot" or single-wire antenna terminal. Avoid shorting the center connection with the shell or receiver will be fully muted.

Speech Synthesizer is an optional circuit that "announces" the tuned frequency through the receiver's speaker. It is quite useful for visually impaired listeners.

Shopping for a Shortwave Receiver

You will probably find looking for a good shortwave receiver a bit more challenging than simply hunting an FM radio or television. Generally speaking, portable shortwave radios made by names familiar to consumers like Sony, Panasonic and Kenwood can often be found in stores selling consumer video and audio equipment. Sales personnel in such retail outlets are seldom familiar with the world of shortwave. Test-tuning a portable receiver using only its pull-up antenna and confined inside a partially metal enclosure is equally useless and typically leaves one discouraged. Possibly the store will have access to an outside antenna for tuning or will demonstrate a unit outdoors for you. Another neighborhood source of portable shortwave radios and "low end" communications receivers are local area Radio Shack stores. Again, checking out performance on the spot depends on congeniality and knowledge of salespeople. Slick talk may be enticing, but it does not increase a particular receiver's sensitivity for receiving weak signals or its selectivity to reject adjacent frequencies.

The more attractive sources of portable shortwave radios, mid-range communications radios, and "high end" shortwave receivers are at communications dealers nationwide. These dealers typically have a knowledgeable sales staff and outdoor antennas ready to connect and demonstrate any and all models on-the-spot for you. They also carry a full line of shortwave magazines and books with up-to-date frequency lists and books. Some of the manufacturer's names of better grade portable shortwave radios include Grundig, Sangean, and Sony. Some of the most well-known names in superb performance communications receivers include Kenwood, Yaesu, Icom, AOR, Lowe and Drake. As a general "familiarizing guide" for shortwave receiver shoppers, brief descriptions of some popular shortwave receivers follow. Each manufacturer produces several models, so remember this is only a sampling of available receivers.

Grundig - This famous German-based manufacturer produces an impressive line of portable receivers for discriminating shortwave broadcast listeners. The Yacht Boy 400 shown in Figure 3-14 is a good representative of Grundig's line. This compact "world band receiver" covers the standard AM and FM bands, tunes in AM stations from 1.7 to 30 MHz, and scans selected bands. Frequencies of interest can also be entered directly via the keypad and stored in the receiver's 40 memories. The Yacht Boy 400 is especially designed for those who travel or lead independent lifestyles. It includes clock radio functions, a pull-up antenna (and a socket for an external antenna), and operates from 6-AA batteries. Performance is quite good and the unit is handy for staying informed about world affairs when traveling.

Figure 3-14 *The Grundig Yacht Boy 400 measures only 5" x 8" x 1.5" HWD, and provides reception of AM international shortwave broadcast bands. It is ideal for staying abreast of world affairs when moving around or traveling.*

Lowe - Shortwave monitoring has always been a popular interest in the United Kingdom and British-made Lowe receivers are widely recognized for their performance. The model HF-150 shown in Figure 3-15

Figure 3-15 The Lowe HF-150 is a British-made all mode receiver with continuous coverage from 3 to 30 MHz. Sensitivity and selectivity are very good. Unit measures 3.2" x 7.2" x 6.5" HWD, and operates from an external battery pack or AC adapter.

vividly illustrates that fact. This compact unit covers 30 kHz to 30 MHz with reception of AM, CW and upper or lower Single Sideband. A synchronous detector to minimize the fading of AM signals is included along with selectable bandwidths for rejecting adjacent channel interference. The HF-150 also features 60 memories that store any frequency and mode and a variable tuning speed that changes according to how slowly you turn the dial. The receiver is enclosed in a rugged wraparound metal case, and an accessory kit containing a pull-up antenna, rechargeable battery pack, carrying handle, and shoulder strap is available for portable operation.

Drake - Shifting focus to American-produced shortwave receivers, we next spotlight R. L. Drake, a widely respected name in affordably-priced communications equipment. Our overview of receivers covers the range from simple to elaborate, so let's move up a couple of steps and highlight the Drake R-8A shown in Figure 3-16. In addition to all-mode continuous coverage from 100 kHz to 30 MHz, the R-8A features superb internal circuitry to pull in weak stations and eliminate adjacent frequency interference. Controls aiding in this respect include passband offset (for minimizing undesired signals), a notch filter (for removing "howls" or tones), an RF preamp (for boosting weak signals), and an RF attenuator (for reducing unusually strong signals). The R-8A also features synchronous AM reception to minimize fading, 440 memories, multiple modes of scanning, an interface for computer control, and much more.

The R-8A could easily be described as a wolf in sheep's clothing. Its outward appearance may not be high glamour, but its internal circuitry is a knockout. The receiver operates on 120 and 240 volts AC or 11 to 16 volts DC. It is suitable for base, field or portable use. Other receivers in the Drake line include a portable and an AM-only unit, and they reflect Drake's pursuit for being tops in their class.

Figure 3-16 The Drake R-8A is a true state of the art communications grade receiver. Unit sports a wealth of features for pulling in weak signals buried by interference. This American-made unit is excellent for monitoring all types of worldwide shortwave communications.

Kenwood - This well-known Japanese manufacturer of home and consumer electronics also produces a full line of shortwave radios and accessories. One of their top receivers, the model R-5000, is shown in Figure 3-17. This communications-grade unit receives all modes and frequencies between 100 kHz and 30 MHz. Many of the communications-grade features such as those included in the Lowe and Drake are also included in the R-5000. One must look more closely to determine actual operating differences. The R-5000's IF Shift is not quite as effective as the R-8A's passband offset,

Figure 3-17 The Kenwood R-5000 is a sharp looking and good performing communications grade receiver with numerous features. Determining whether it is slightly better or not quite equal to similar units depends on personal needs and may require a side by side off-air monitoring comparison.

for example, and its notch filter cannot correct for reduced sensitivity when removing "howls." The R-5000, however, includes selectable bandwidths, dual noise blankers, 100 "store everything" memories and many more features. It is an attractive and cost-effective unit capable of filling almost any need in fine style. The R-5000 may be powered from a 120 volt AC or 12 volt DC source.

Antennas for Shortwave Reception

The ultimate performance of any shortwave receiver large or small, lavish or inexpensive, is directly influenced by its accompanied antenna. Simply stated, it is the radio's lifeline to the outside world! Portable radios often feature a built-in antenna, but their small size limits performance. Antennas work best when made for your specific bands of interest or homemade and cut to length for your selected band(s). If you want to listen using several bands, a multiband antenna is a very good investment. The question of whether one should buy or build their own shortwave antenna depends solely on individual preferences. Will a homemade antenna work as well as a commercially-made equivalent? Sure, provided you take care building it and weatherproofing its connections. One popular variety of commercially-made shortwave antennas which can be installed almost anywhere is shown in Figure 3-18. This particular antenna covers all the international broadcast bands. The antenna is also close enough to resonance (optimum state of tuning for a particular frequency) that it is actually good for receiving throughout the 500-kHz to 30-MHz range. As in shortwave receivers, an endless selection of shortwave antennas and parts for making your own antennas are readily available through shortwave and amateur radio dealers nationwide.

Now let's shift focus and discuss building your own antenna at home. We will start with a few basic facts to set your mind at ease. First, any type of wire antenna works best when erected outdoors and installed as high as possible. In most cases, this equates to a height of about 15 to 25 feet at one end and

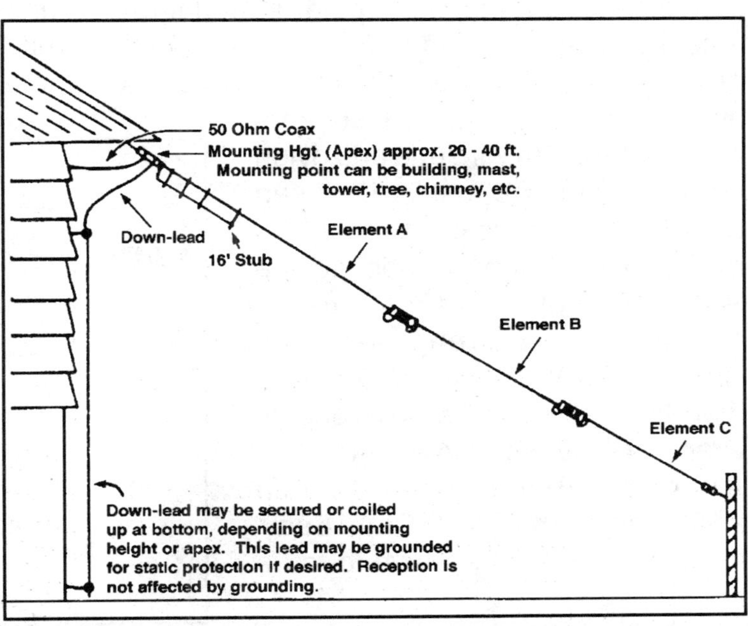

Figure 3-18 Commercially available/preassembled shortwave antennas like the Alpha Delta antenna shown here are readily available from dealers nationwide.

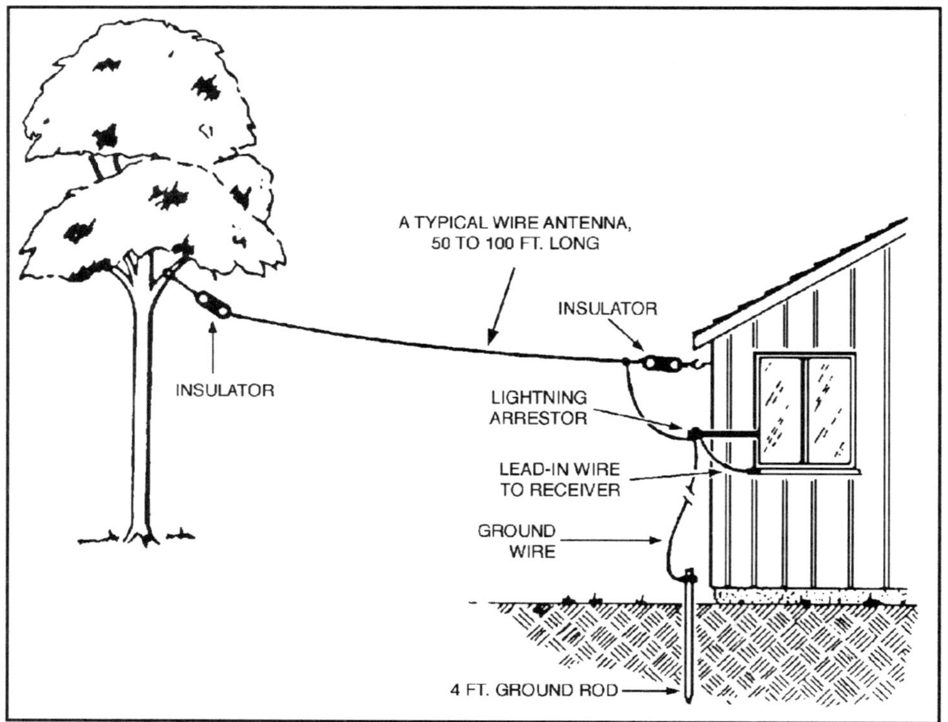

Figure 3-19 *A typical long-wire antenna. Exact length is not a critical factor.*

35 to 50 feet at the other end. As long as your antenna wire is not the highest object around or low enough to the ground to be a danger to people, it will be fine. Insulated or bare copper wire is the most popular material for antennas. Size or diameter is not important provided the wire does not break under its own weight. Use ceramic insulators at the ends of the wire and add nylon rope between the insulators' open ends and supporting tree limbs or roof edges. When connecting the lead-in wire or coaxial wire to the antenna, make strong and solid splices with bare wire and solder them if at all possible. Thoroughly wrap each connection and junction with electrical tape pulled tight, then add a liberal amount of Coax Seal® for weatherproofing. Strive to route the lead-in wire or coaxial cable at right angles away from the antenna and away from metal objects, yet on a fairly direct path to your receiver. If a bare lead-in wire is used, be certain to support it with ceramic, stand-off insulators and avoid contact with other metal surfaces.

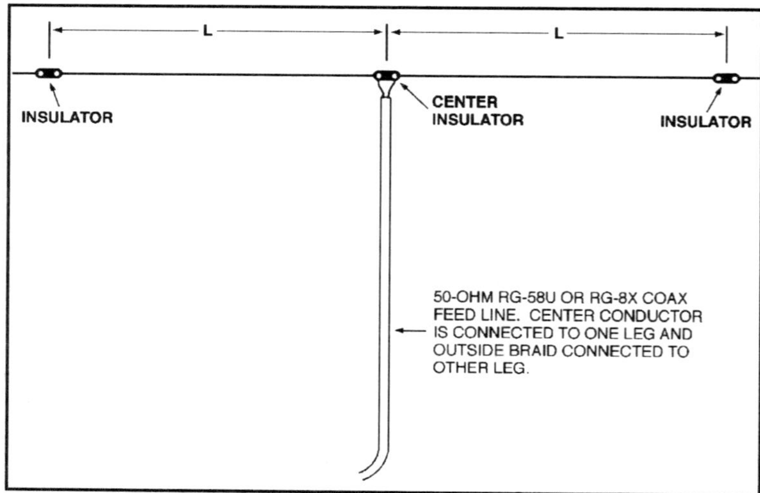

Figure 3-20 *A single-band dipole. For best results, position the antenna broadside to the desired direction of reception.*

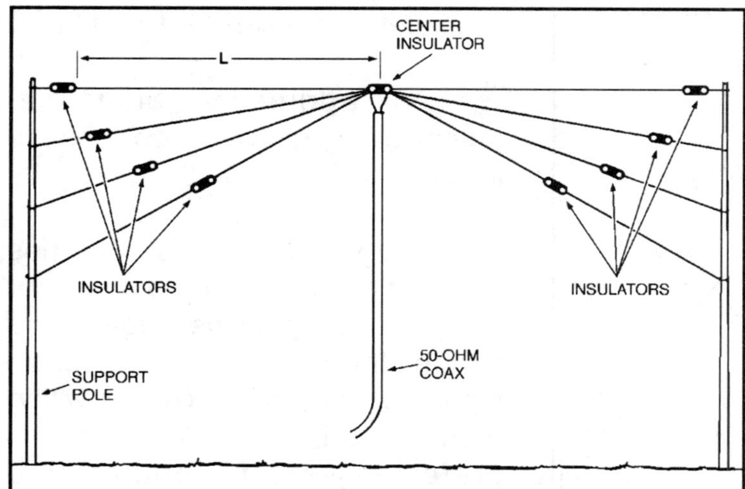

Figure 3-21 *Additional dipoles connected to the center insulator comprise a multiband dipole with a single 50 ohm coax feedline.*

Long-Wire Antennas

One of the simplest, yet effective receiving antennas, is a random long-wire antenna. This antenna is just a length of wire 30 to 100 feet long. It is erected in a straight line on a horizontal plane with an insulated lead routed into your house and then to your shortwave receiver. The long-wire antenna performs best when erected from 20 to 50 feet off the ground; in a spot clear of trees, buildings, and other obstructions; and mounted away from power lines and other areas that could induce electrical noise into your antenna and on into your receiver (Figure 3-19). Both ends of the long-wire antenna must have suitable insulators. The actual antenna wire can be any suitable wire, insulated or bare. The lead-in wire should be insulated to keep it from shorting to other obstructions such as parts of the house, window frames, metal screens, and the like.

The random long-wire antenna, when used with a simple antenna tuner, will cover a frequency range from 1.6 MHz to 30 MHz. Most of the modern solid-state shortwave receivers cover a wide range of frequencies; therefore, we need an antenna that will afford good reception from the low end (1.6 MHz) to the highest end of the receiver's frequency range (30 MHz). The long-wire antenna fills this need for extreme frequency ranges, requiring only a quick peaking of the knobs of a simple antenna tuner. The receiver's "S" meter or signal strength meter is used for maximum signal peaking—just tune for a maximum signal reading on the "S" meter. This maximum tuning takes only a second or two and results in good reception. An antenna tuner is a simple device that will allow you to maximize both the pickup and transfer of the signal from the antenna to your receiver. It consists of a tapped coil and a variable capacitor, both of which are adjusted by knobs

Band	Frequencies (MHz)	Each Side of Dipole (L)	Overall Dipole Length
11 Meter	25.4 to 26.1	9 ft. 1 in.	18 ft. 2 in.
13 Meter	21.45 to 21.75	10 ft. 10 in.	21 ft. 8 in.
16 Meter	17.7 to 17.9	13 ft. 2 in.	26 ft. 4 in.
25 Meter	11.6 to 12.0	19 ft. 10 in.	39 ft. 8 in.
31 Meter	9.2 to 9.7	24 ft. 9 in.	49 ft. 6 in.
41 Meter	7.1 to 7.4	32 ft. 3 in.	64 ft. 6 in.
49 Meter	5.9 to 6.4	38 ft. 6 in.	77 ft. 0 in.
60 Meter	4.75 to 5.0	48 ft. 0 in.	96 ft. 0 in.
90 Meter	3.2 to 3.4	70 ft. 11 in.	141 ft. 10 in.

Figure 3-22 *Dimensions for assembling a dipole antenna at home.*

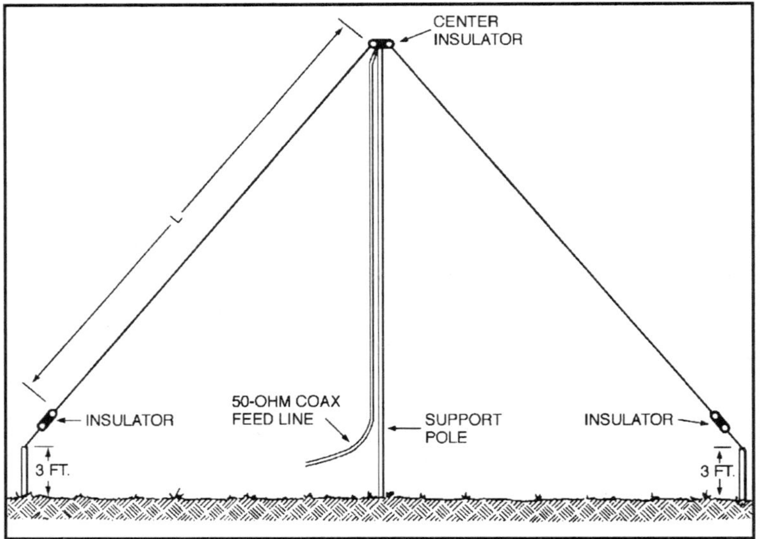

Figure 3-23 *Inverted V antenna. Item is easy to assemble and erect, as it requires only a single center support. Multiband dipoles can also be erected in inverted V configuration.*

on the tuner's front panel. The lead-in wire from your antenna connects to the tuner's input; then a short length of cable connects between the tuner and your receiver's antenna socket. Quick, easy and simple!

Dipoles

The dipole is a basic half-wave antenna that is cut to a specific band of frequencies in the HF spectrum (Figure 3-20). Dipoles and multidipoles are used by many SWL'ers whose principal activity is confined to a single shortwave band or bands. More than one dipole can be combined in the dipole system (see Figure 3-21). Dipole length is calculated by the following simple formula: $\frac{234}{L} = F$ where:

L = length in feet for each half of the dipole,
F = center frequency of the band in MHz.

For example, the center frequency of the 25-meter broadcast band (11.6 MHz to 12.0 MHz) is 11.8 MHz.

$$234 \div 11.8 \text{ MHz} = 19.83 \text{ feet}$$

As additional guidance, precalculated dipole dimensions are included in Figure 3-22.

The single dipole, or a series of multidipoles, will have a directivity factor broadside to the length of the dipole; therefore, a cut or tuned dipole can be erected to favor, within a preferred broadcast band, a preferred set of stations or a particular area of the world.

Inverted-V Antennas

The inverted-V is basically a dipole mounted in a V position. This has the advantage of saving space, and it requires only a single center pole for the main support (see Figure 3-23). A low-cost center support can be made easily using regular TV mast sections up to 25- to 30-foot heights.

Figure 3-24 *Active antenna system made by Palomar Engineers measures only 7 x 6 x 2", and provides good reception when erecting an outdoor antenna is not feasible.*

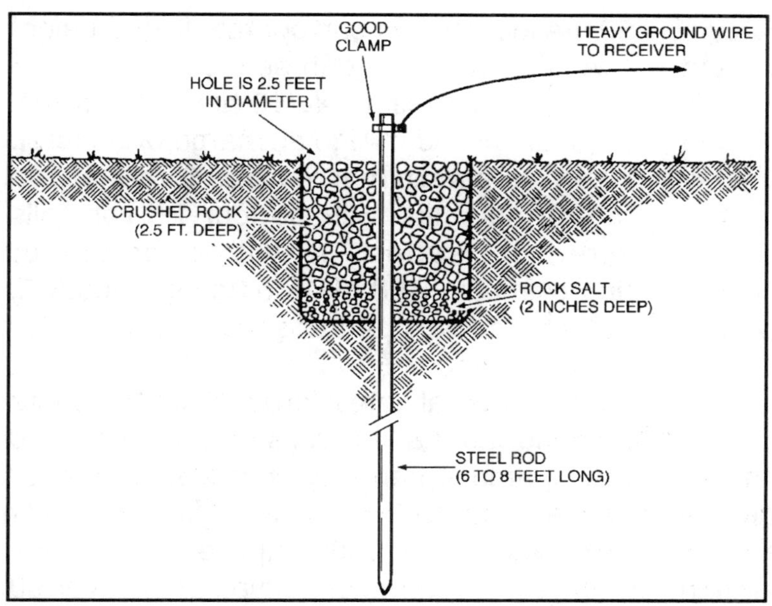

Figure 3-25 *A good ground system should be installed as close as possible to the shortwave receiver.*

Ground Systems

A simple and effective RF ground system is shown in Figure 3-25. It consists of a steel, copper-clad ground rod, 6 to 8 feet long, driven into moist ground. The clamp must be securely attached to obtain a proper ground-wire connection. The ground wire must be heavy and as short as possible. In dry climates and dry periods of the year, a generous watering of the ground-rod area will do wonders for your ground system. Alternately, connecting a wire to your outdoor cold water pipe will serve as an adequate ground system.

Alternate Antenna Systems

When all else fails and you cannot come up with a good outside antenna, and you do not feel a makeshift indoor antenna will do, there is one other good possibility left! This would be one of the newer indoor active antennas that are available from most shortwave dealers. This system is a completely self-contained active device with electronic sections to tune all the major shortwave bands. The units have an attached whip antenna and a tunable preselector with sufficient amplification to deliver a good signal to the receiver. The indoor units sit on your desk—ready to listen to the world. Unique circuitry minimizes intermodulation and provides added RF selectivity. These units can also be used as preselectors for external or outside antennas. Most active antenna systems cover frequencies of 300 kHz to 30 MHz in 5 bands. The units are simple to hook up and use. An example of an active antenna system is shown in Figure 3-24.

Power Sources for Shortwave Receivers

Simple shortwave receivers can usually be powered by batteries, and an AC adapter is usually available for extended home use. Larger communications receivers, however, may be limited to only 120-volt AC power. This is fine provided AC power is available, but AC power is usually the first to go during an emergency. Receivers capable of operation on 12 volt DC truly have their advantages. As we think ahead about our shortwave receiver's application, we may realize battery power will be necessary for field use during survival situations. A good all-around solution may be purchasing an inexpensive shortwave receiver for casual listening and as a backup unit, but purchasing a more sophisticated communications receiver for more serious applications—one which is lightweight and capable of operating from 12-volts DC or 120-volts AC.

Generally speaking, regular carbon batteries yield the shortest operating time and shelf life. These inexpensive cells, however, are readily available at grocery stores and convenience stores on almost any corner—which is an advantage at times. A better choice is alkaline batteries, because they have a greater ampere-hour rating and longer shelf life. Alkaline batteries will keep your shortwave radio running two to three times longer than regular carbon batteries.

Assuming you have a charger and prefer the convenience of portability, rechargeable nickel cadmium batteries are a good choice. Their shelf life between charges is less than alkaline batteries but slightly greater than carbon batteries. Two additional good choices are gel cells and sealed lead acid batteries. You will probably need to make an adapter or use clip-lead jumper cables to connect them to your radio. The beauty of gel cells and sealed lead acid batteries is they do not tend to build a "recharge memory" in the way nickel cadmium batteries will if they are not completely discharged before recharging. In other words, you can use them for a brief period, then "retop them" to full charge when needed without concern for damaging the battery.

All radios do not require the same kind of batteries or same voltages for operation (a point to keep in mind when planning your backup power system). Small "AAA" and "AA" cells deliver 1.5 volts each when fully charged (exception: nickel cadmium batteries usually deliver 1.3 volts when fully charged). The smaller cells are usually rated at .5 amperes, or 500 milliamps (mA). By studying your receiver's manual and technical specifications, you can determine how much current is required for each hour's use. For example, a radio that is rated at 3 volts and 50 mA input power can operate from a pair of "AA" batteries (wired in series to produce 3 volts) for approximately 10 hours (50 mA X 10 hours = 500 mA: total charge of battery). Likewise, a radio that requires 6 volts at 100 mA for operation will use four 1.5 volt cells wired in series. It will operate for approximately 100 mA X 5 hours = 500 mA, or 5 hours total operating time.

Additional notes: large "C" cell batteries are also rated at 1.5 volts and deliver 1,000 mA (1 amp) when fully charged. Larger "D" cells are rated at 1.5 volts and typically deliver 1250 mA (or 1.250 ampere) when fully charged. 9-volt batteries typically deliver 90 mA, or .09 amperes. 12-volt gel cells and sealed lead acid batteries are usually stamped with their voltage and current rating. A 12-volt battery approximately 3 x 6 x 4 inches will probably deliver 5 to 6 amperes of current. If your shortwave radio requires 12 volts for operation and uses 100 mA of current, you can expect a fully charged 12-volt gel cell or lead acid battery to power it for approximately 100 mA per hour x 24 hours/day, or 48 to 50 hours. Remember that cells in series add voltage and their current (mAh) stays the same. Cells wired in parallel deliver the same voltage as a single cell, but their mA rating is multiplied by the number of cells in the battery or bank of cells.

The benefits of shortwave monitoring are truly endless and priceless. During times of uncertainty, it helps clear distortions surrounding political or economic unrest and confusions in world affairs. During calm times, it is a superb educational tool and a front row seat to world cultures—and much more. We heartily encourage you to investigate the many attractions in shortwave in the near future.

CHAPTER FOUR

The World of Amateur Radio

Whether attracted by the aspects of emergency preparedness or capabilities of global communications, becoming a licensed Amateur Radio operator holds significant merit for all survival-oriented individuals. This fact is particularly true for people who live independent lifestyles or reside in remote areas, and who desire a viable link with the outside world. Their need for knowledge or assistance can occur anytime, anywhere. Amateur Radio has many unique benefits not found in other communications services. It is friendly, person-to-person communications with no charge for each minute's use; and it is recognized as the world's greatest electronic related hobby. This does not mean you need to be a technical "whiz" to pass an Amateur Radio licensing exam and set up your own equipment. Folks from 8 to 80, including older men (OM in ham lingo) and young ladies (YLs in ham terms) do so all the time, and so can you.

Becoming an amateur radio ("ham") operator offers even more favorable rewards, many of which easily captivate one's interest for at least two lifetimes. Radio amateurs enjoy weekend contests, exchanging quick words with amateurs in other countries to share knowledge, communicating via satellite and ham TV, and even bouncing signals off the moon. Amateurs also exchange printed messages via Radio Teletype (RTTY) or wireless computer-to-computer links called Packet radio.

Figure 4-1 *Amateur Radio setups need not be large or elaborate to be effective. A conventional HF installation as shown here is capable of globe-spanning communications any hour of the day or night. It also tunes/receives all shortwave broadcasts, private and commercial communications worldwide, and serves as an information center during emergencies.*

Figure 4-2 *Ultra compact HF amateur radio transceivers and screw-apart mobile/portable antennas provide globe-spanning communications from anywhere in the world.*

In fact, amateur radio setups are found everywhere from military airplanes flying thousands of feet above the earth, to private homes, autos, boats, motorcycles and bicycles. Portable and battery-powered setups are popular when traveling, vacationing, and camping. Many Civil Defense and military facilities also have amateur radio setups, and ham operators often communicate with them during times of emergency situations. The prime focus of amateur radio is emergency communications, so amateurs maintain close contact with Civil Defense, Red Cross, Federal Emergency Management Agencies (FEMA) and Emergency Operating Centers. Two additional emergency services within amateur radio itself are the Radio Amateur Civil Emergency Service (RACES) and Amateur Radio Emergency Corp. (AREC), both of which practice emergency operations with battery and portable power systems on a daily, weekly, or monthly basis. During emergencies, the groups join together and form a network using normal amateur radio band frequencies and relay messages between local, state and federal officials.

People occasionally attempt to visualize amateur radio as CB because both are hobbyist-type communications, but the similarity stops there. A license is not required for CB. A valid license and F.C.C. issued call sign are absolute necessities for transmitting with an amateur radio setup. As mentioned in previous chapters, a license is not required to simply listen on any frequency. Rather than being confined to a single band, radio amateurs are authorized to use a variety of bands within the HF, VHF, UHF and microwave spectrum (HF, VHF, UHF are the initials for High Frequency, Very High Frequency, and Ultra High Frequency bands). Amateurs are also allowed much greater power than CB. A maximum of 1500 watts is permitted on the HF bands. Finally, a higher level of discipline and refined on-the-air operating techniques are immediately obvious on any and all amateur radio bands. A simple example involves the highly popular use of range-extending FM repeaters on the 2-meter band. Several people within a group will listen/receive while only one person talks/transmits at a time. After each operator talks or comments (and switches back to receive), the repeater emits a courtesy "beep." The next

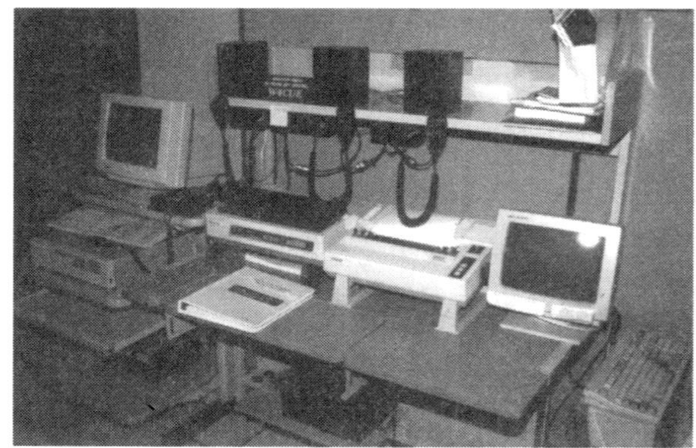

Figure 4-3 *Impromptu, yet effective, amateur radio setups large and small are found in all types of emergency situations and public service facilities.*

Figure 4-4 *An easy to obtain Technician Class License gives one direct access to all hot and thriving VHF and UHF Amateur Radio bands used for local area communications. Long-range communications via 6 meters, amateur satellites, and packet radio links are also authorized. This introductory license is a priceless asset during any urgent or emergency situation.*

operator will typically pause a second or two, perchance someone else needs to speak, then begin a follow-up transmission. Refined techniques are apparent on the HF amateur bands, as each station typically begins and ends each transmission with related parties' call signs. Not only do those letters and numbers correspond to each station's location, they also mark the start and stop points of transmissions to avoid confusion from both parties talking at once.

You may have heard that in addition to passing an examination on radio practice and theory, learning the Morse code is necessary for obtaining a ham license. A decade ago, that was true. Today, however, amateur radio's introductory (Technician class) license does not require any knowledge of the Morse code. Only a simple exam on operating practices, F.C.C. rules, and basic electronic theory is required. In return, a licensee can operate through local area FM repeaters, communicate via satellite, via amateur TV, and several other modes in the 2-meter, 6-meter, and higher frequency amateur radio bands. Combine those privileges with the fact that repeaters are located in large and small cities throughout the country, indeed the world! In addition, many of the repeaters include autopatching and direct telephone accessing capabilities; and the Technician class license becomes quite appealing.

Operating within the globe-spanning HF (high frequency) amateur bands requires passing a slightly more complex examination and demonstrating proficiency in copying Morse code at 5 words per minute for 10-meter privileges or 13 words per minute for operating privileges on lower bands such as 20, 40, 30, 80, 17, 12 and 160 meters.

Another misconception is that amateur radio equipment is expensive, complicated, and interferes with neighbors TVs. Again, things have changed in the last decade! Newer amateur radio transceivers emit exceptionally "clean" and well-filtered signals that seldom affect other electronic equipment. Ham transceivers have become more complex, true; however, new equipment is extremely reliable and seldom requires major repairs. Further, you do not need to install special wiring for an amateur radio setup. A typical 100-watt HF transceiver plugs into a regular home 110-volt AC outlet. If a super powerful amplifier is used, it can plug into a 220-volt AC outlet like those used for an air conditioner or clothes dryer.

Figure 4-5 Compact HF transceivers fit into limited spaces and provide globe-spanning communications and full shortwave reception right from the auto.

Expensive? Some new Technician-class licensees have started out with a used 2-meter handheld "talkie" purchased at a hamfest for less than $50. Some higher class licensees have started out on the HF bands with a 40- or 50-watt setup purchased for $100 (and enjoyed many long distance contacts!). More typically, however, an average new 2-meter talkie is in the $300 price bracket and a "first-rate" HF transceiver with 100 watts output and capable of worldwide communications is in the $1,000 bracket.

A final misconception we can "clear up" is Amateur Radio is only for those with electronics backgrounds. In fact, Amateur Radio has no confinement regarding backgrounds. Ham operators include lawyers, doctors, paperboys, housewives, professional entertainers, kings, and folks just like you and me. Some well-known amateurs include entertainer Ronny Milsap, WB4KCG; news broadcasters Walter Cronkite, KB2GSD and Roy Neal, K6DUE; King Hussein of Jordan, JY1; Air Force General Curtis Le May, W6EZV; Senator Barry Goldwater, K7UGA; and astronaut Owen Garriott, W5LFL.

Amateur radio can be visualized in two general categories: that used for long range/global communications (HF), and that used for local area communications (VHF/UHF). As you will recall from our previous chapters, the HF spectrum between 1.7 and 30 MHz has the unique capability of being reflected by the earth's ionosphere (and double or triple skipping) to reach around the world. The spectrum above HF, which is called VHF, UHF and microwaves is used for (electronic) line-of-sight communications. The exceptions to that rule are satellites and earth-based repeaters on VHF and UHF bands that extend the communications range. How far? That depends on how far the relay station can "see." If the relay station is on a high mountain, it may extend the communications range 200 miles. If the signal relaying unit is aboard a satellite 20,000 miles above the earth, it may extend the communications range over one or two continents. Now let's take a closer look at each of these amateur radio worlds.

The HF World of Amateur Radio

This is the original or "classic" area of amateur radio that has been its heartbeat of activity since radio's beginning in the early 1900's. It is used for communicating worldwide, within the U.S., and in local areas. Today's HF amateur setups typically fit on a medium-size desk or into a modified closet rather than completely filling a garage or

Figure 4-6 *License plates with Amateur Radio call letters and unique antennas signify mobile ham radio setups.*

basement with large equipment. When driving through cities or neighborhoods, you can occasionally spot homes of amateurs with HF stations by their tall towers and rotatable "beam" antennas. Some HF setups in antenna-restricted areas or connected to budget-conscious stations, however, use inconspicuous wire antennas or even indoor antennas. Vertical antennas and numerous types of disguised or "hidden" antennas are also used for HF bands operation. Mobile HF setups capable of spanning the globe right from an auto are also quite popular. They usually can be spotted by the larger-than-CB antennas on autos and occasionally by ham call letters on vehicle tags.

A typical medium-budget home amateur radio setup consists of a 100-watt transceiver with microphone, telegraph key, and maybe a computer for data communications such as Packet or RTTY. Popular antennas for such setups are verticals in the 20 to 40 foot high category and wire (dipole) antennas between 33 and 133 feet in length. Average cost of the overall setup is between $400 and $1500.

A more elaborate setup typically consists of a slightly larger deluxe transceiver, microphone, key, home computer, and high power amplifier in the 1,000 watt to 1,500 watt category. These setups are usually connected to tall towers with larger beam antennas and usually more than one antenna for various bands. The average price is $5,000 to $15,000. Both economical and elaborate stations are capable of spanning the same distance. High power stations simply "stand above the crowd" with a booming signal when "skip" conditions are less than ideal. A typical mobile station consists of a small, yet impressive in its performance, 100-watt transceiver (similar to a home transceiver) connected to a 8- to 12-foot antenna on the vehicle. These setups also have globe-spanning range. However, their small antenna usually makes their signal slightly weaker than home stations under less than ideal band conditions.

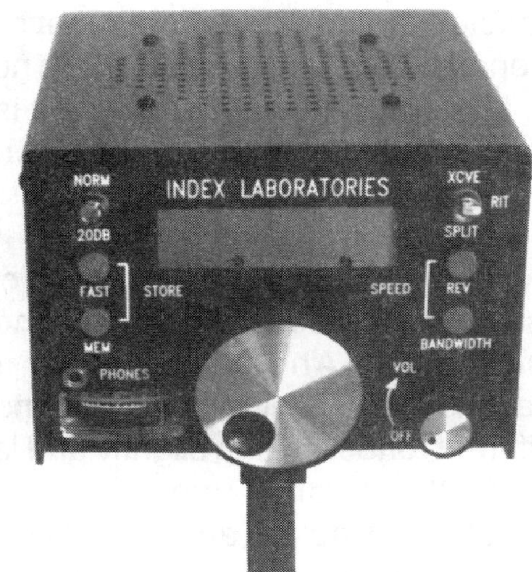

Figure 4-7 *Incredibly small and low power (QRP) HF Amateur Radio transceivers like the Index Labs "QRP Plus" shown here goes anywhere, operates for many hours from batteries, and reaches around the world when band conditions are good.*

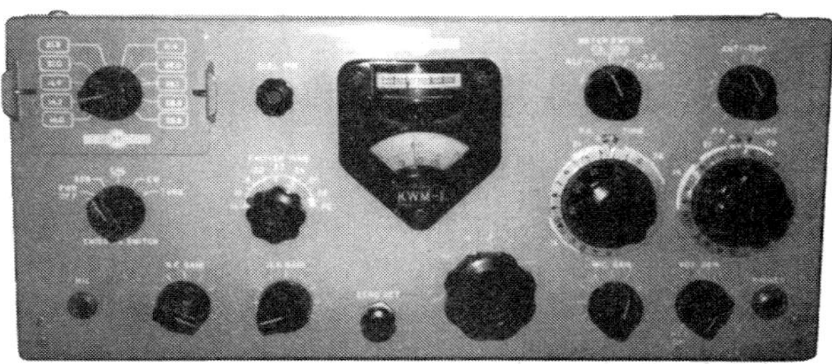

Figure 4-8 Older model vacuum tube HF transceivers are popular among many survival-oriented amateurs because they can withstand greater levels of EMF than solid state equipment.

Additional special purpose HF setups also warrant mention. An increasing number of radio amateurs are enjoying long distance communications using very low power levels, called QRP in ham terms (5 watts or less). Such setups are not fully reliable under all band conditions, but do have the advantage of being powered by batteries and/or solar energy panels for long periods of time. Some amateurs also use older model transceivers with vacuum tubes rather than solid state devices and transistors. In addition to the romance of soft glowing tubes and beautiful sounding audio, vacuum tubes are much more hearty and capable of surviving a reasonable amount of electromagnetic pulse (emp) energy like that associated with nuclear blasts. Modern solid state equipment is very good and quite reliable, but a distant burst of emp could destroy it.

Amateur radio voice communications on HF use Single Sideband (voice) and CW (Morse code). Another popular communications mode, radio teletype, sounds like two different tones shifting slightly and lasting between three and five minutes per transmission. Packet and AMTOR communications sound similar to crickets chirping. They are typically short bursts, then the other "cricket" (station) "chirps" back. Finally, slow scan television sounds like musical tones with occasional "bleeps." These video pictures are transmitted in the SSB portions of amateur HF bands. Now let's briefly look at each of the bands and their ranges (See Figure 4-9).

160 Meters is located only slightly above the standard AM broadcast band. 160 meters covers from 1800 kHz/1.8 MHz to 2000 kHz/2.0 MHz. A General or higher class license is required for use. By a gentleman's agreement, the lower 15 kHz section is allocated for CW and the upper range is used for SSB. This is a medium-range late evening band, typically "opening" a couple of hours after sunset and "closing" around sunrise. The communications range varies from local area to 300 or 400 miles, at times extending nationwide, and sometimes reaching South America and Europe during wee hours of darkness. This is the oldest and most famous amateur band, the atmosphere is usually casual like a country store, and related 160-meter antennas are usually very long wires (130 to 260 feet).

80 Meters This is another nighttime band similar to 160 meters, but supporting a wider variety of activities. The CW segment extends from 3.500 MHz to 3.750 MHz, and the SSB portion extends from 3.750 MHz to 4.0 MHz. Within the CW portion, you may also hear RTTY and Packet signals. Within the SSB portion, you may also hear SSTV (Slow Scan Television) signals. A large number of state, regional and emergency nets meet daily and weekly within the SSB (Single Side Band) portion. Information directly from hurricane and other emergency related areas is often heard. When looking for specific information regarding emergency situations, check various conversations only briefly to determine if you have tuned in an appropriate net. If not, continue tuning for more information. Antennas for this band are also typically wires 60 to 130 feet in length.

AMATEUR HF (HIGH FREQUENCY) BANDS CHART

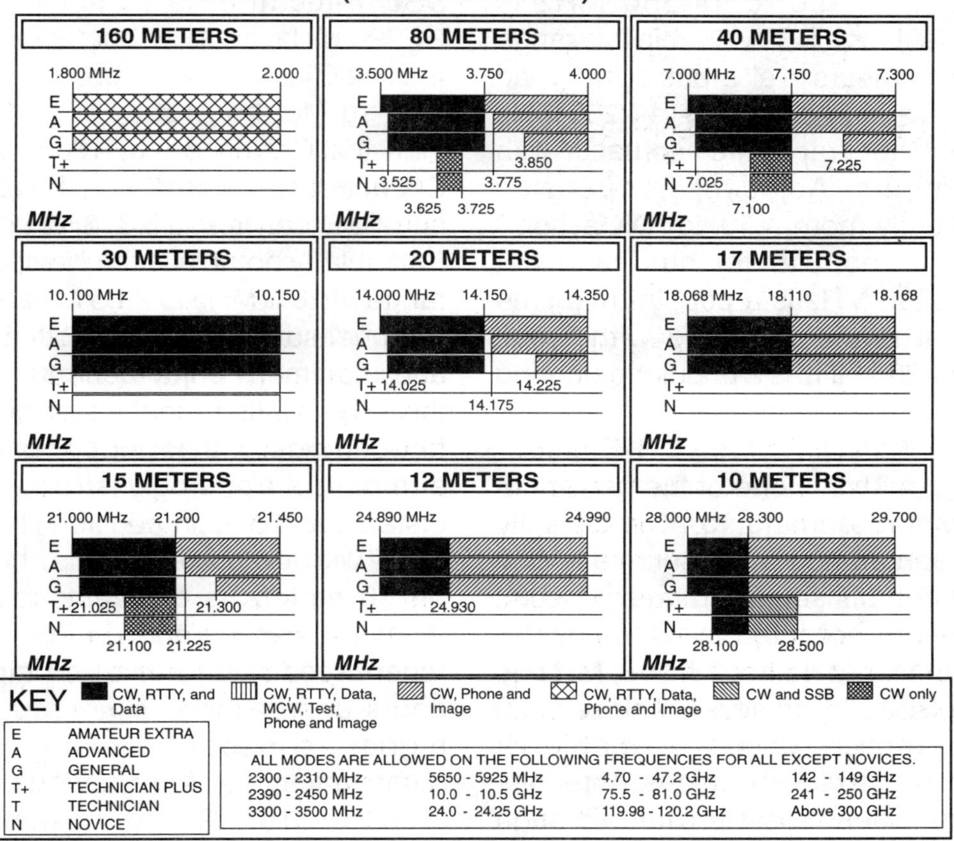

Figure 4-9 *Outline of the Amateur Radio HF bands categorized according to frequency in kHz/MHz and operating/transmitting range authorized for each class of license.*

40 Meters Although primarily a nighttime band, 40 meters is often "open" all hours except for a lull between 10 a.m. and 2 p.m. in local time zones. The frequency range of 7.0 MHz to 7.150 MHz is allocated for CW, and the range of 7.150 MHz to 7.300 MHz is allocated for SSB and SSTV. This is one of the more popular amateur bands for evening use, and its SSB portion is also shared with some foreign broadcast stations. Slower speed and easier to copy novice Morse code activities will be noted between 7.100 MHz and 7.150 MHz. Antennas for this band are also usually wires approximately 67 feet in length, or verticals between 17 and 35 feet in height.

30 Meters This band is often open around the clock, except for a slight lull between 11 a.m. and 1 or 2 p.m. local time. It is a CW-only band within the 10.100-MHz to 10.150-MHz range. Operation is limited to General or higher class licenses and activity is presently rather light. This is also the only amateur radio HF band with a maximum power limit of 200 watts in the U.S. As a result, it is a true "sleeper." Many low power and foreign stations frequent this band, and contacting them is quite easy, provided one is proficient in Morse code.

20 Meters Truly the most popular amateur radio band for both U.S. and global communications on a 24-hour a day basis. Indeed, finding a blank frequency on this band during weekends is nigh impossible! The range of 14.000 MHz to 14.150 MHz is allocated for CW and RTTY/Packet/AMTOR.

The range of 14.150 to 14.350 MHz is allocated for SSB and SSTV. Within this range you will hear amateurs in all areas of the world conversing in various languages. 14.200 MHz is a "hot spot" for "rapid fire" contacts with distant amateurs. Adjacent frequencies support a slightly more relaxed "DX action." 14.313 MHz is popular for hurricane net activities, 14.300 MHz is a popular maritime frequency for communicating with small boats, and 14.303 MHz is a popular gathering spot for HF mobile enthusiasts.

17 Meters This is one of the newer HF bands available to amateurs. It is typically "open" between an hour after sunrise to one or two hours after sunset. The range of 18.068 to 18.110 MHz is allocated for CW, and the range of 18.110 MHz to 18.168 MHz is allocated for SSB. Being situated close to 20 meters, 17 meters is also capable of supporting global communications. Antennas for 17 meters are not as plentiful (although easily homemade, as they are only 25 feet in length), consequently activity has thus far been rather light. As a result, lower power stations "stand up" very well on 17 meters.

15 Meters This is primarily a daytime band capable of supporting global communications between sunrise and sunset during years of high sunspot activity. The lower segment of 21.0 MHz to 21.2 MHz is allocated for CW, and the upper segment of 21.2 MHz to 21.450 MHz is allocated for SSB (and SSTV). Antennas for 15 meters are also small (typically 22 feet for wires), and "beam" antennas similar to very large-sized TV antennas are quite popular for this band and the two upper bands.

12 Meters This is also a lightly used and newer band good for daytime communications on a long-range basis. The CW segment is 24.890 MHz to 24.930 MHz. The SSB allocation is from 24.930 MHz to 24.990 MHz.

10 Meters Slightly higher in frequency than the Citizen Band, 10 meters supports CW activity between 28.0 MHz and 28.3 MHz, and SSB between 28.3 MHz and 29.7 MHz. Particularly notable is the Novice/"Tech Plus" range of 28.3 MHz to 28.5 MHz. During years of higher sunspot activity, both new licensees and old-timers enjoy communicating within this window. Its range is typically nationwide to worldwide, with lower power signals often pumping surprisingly strong signals into distant locations. In addition to spanning the globe with low power signals, the big surprise on 10 meters is the vast amount of new amateur licensee activity. The band is often "open" soon after sunrise and into the evening hours approaching 9 p.m. during years of higher sunspot activity. Emergency communications from various sources are found on 15, 12 and 10 meters—mainly because these bands are more oriented toward person-to-person, long distance communications than nets. When tuning in SSB signals on any of the amateur HF bands, the keynote is patience and slow tuning. Turn your receiver's dial as if you are "cracking a safe." A bit of expertise is also required for tuning SSB signals. With your hand on the dial, make adjustments as the other person talks and stop adjusting when they stop talking. Experience will guide you to turn the tuning dial right or left according to whether the transmitting station is tinny or bassy.

The VHF/UHF World of Amateur Radio

This is the newer area of amateur radio, and it is a literal hotbed of local area FM mode activity. Indeed, one can travel coast to coast

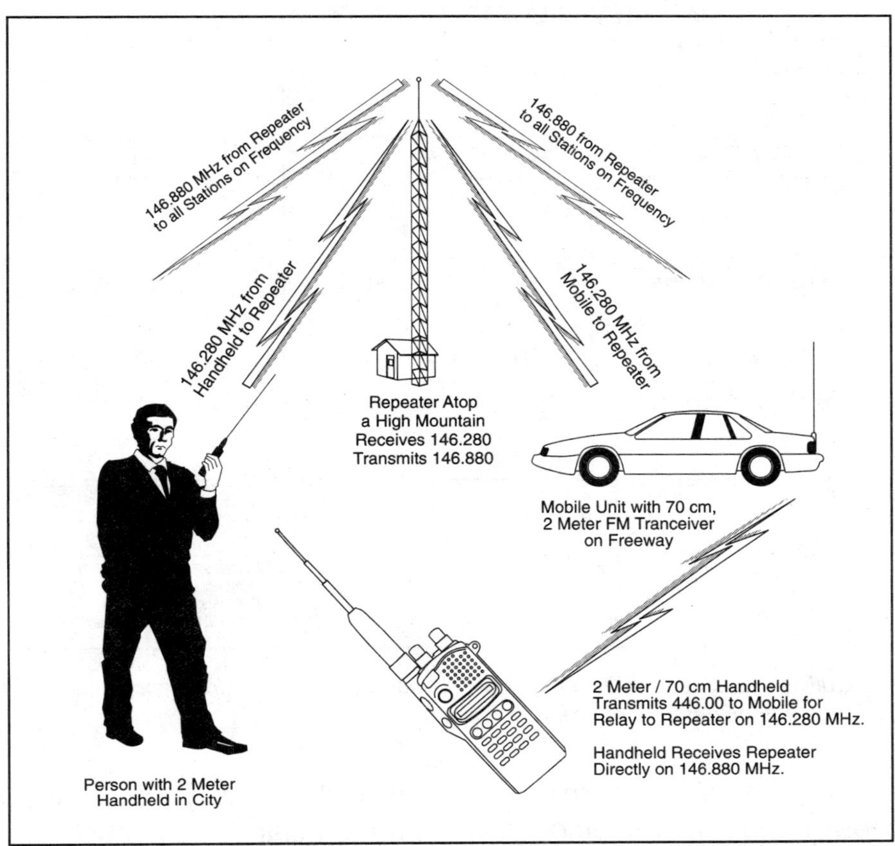

Figure 4-10 *High power repeaters atop tall buildings or mountains significantly extend the range of base, handheld and mobile FM signals by simultaneously receiving and transmitting on separate frequencies. Several new style VHF/UHF mobile transceivers are also equipped to crossband-repeat signals between an ultra-small shirt pocket "talkie" used in inaccessible areas and a distant repeater.*

monitoring various amateur radio 2-meter and 70-cm repeaters along the way and seldom be out of coverage range. The vast majority of amateurs new and old have at least one 2-meter/70-cm handheld and/or mobile FM transceiver. If only half of them turned on their radios at the same time, they could easily relay a voice message "hand off style" across the country. Needless to say, the flexibility, portability and simplicity of operating these VHF/UHF bands—plus the lower cost of equipment and attractions of emergency preparedness—are extremely appealing to new licensees.

A new amateur usually starts out with a handheld FM transceiver that runs two to five watts of output. A mobile FM transceiver running between 10 and 70 watts output (plus a mobile antenna) is typically added soon afterwards for convenience and maximum range on the open road. A bit later, the new amateur may connect his home computer to the mobile transceiver (moved into the house) for data communications via Packet radio. The amateur may also add a multi-mode transceiver (FM, SSB, and CW) to the setup for working through amateur radio satellites and/or communicating with the space shuttles during flight and orbiting space stations such as MIR. Power levels between two watts and 50 watts are common on VHF/UHF, because they get a "boost" through repeaters in various areas.

AMATEUR VHF/UHF FREQUENCY BANDS CHART

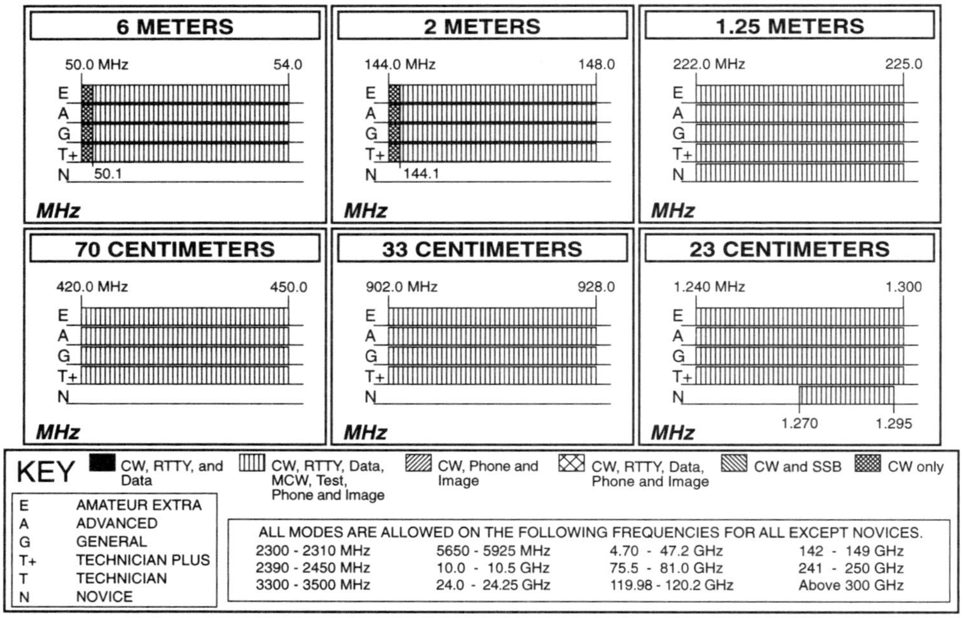

6 METERS
50.0 MHz 54.0
E A G T+ N 50.1
MHz

2 METERS
144.0 MHz 148.0
E A G T+ N 144.1
MHz

1.25 METERS
222.0 MHz 225.0
E A G T+ N
MHz

70 CENTIMETERS
420.0 MHz 450.0
E A G T+ N
MHz

33 CENTIMETERS
902.0 MHz 928.0
E A G T+ N
MHz

23 CENTIMETERS
1.240 MHz 1.300
E A G T+ N
1.270 1.295
MHz

KEY: CW, RTTY, and Data; CW, RTTY, Data, MCW, Test, Phone and Image; CW, Phone and Image; CW, RTTY, Data, Phone and Image; CW and SSB; CW only

E	AMATEUR EXTRA
A	ADVANCED
G	GENERAL
T+	TECHNICIAN PLUS
T	TECHNICIAN
N	NOVICE

ALL MODES ARE ALLOWED ON THE FOLLOWING FREQUENCIES FOR ALL EXCEPT NOVICES.

2300 - 2310 MHz	5650 - 5925 MHz	4.70 - 47.2 GHz	142 - 149 GHz
2390 - 2450 MHz	10.0 - 10.5 GHz	75.5 - 81.0 GHz	241 - 250 GHz
3300 - 3500 MHz	24.0 - 24.25 GHz	119.98 - 120.2 GHz	Above 300 GHz

Figure 4-11 *Outline of the Amateur Radio VHF/UHF bands categorized by frequency in MHz and operating/transmitting range authorized for each class of license.*

Simply explained, a repeater is a combination receiver and transmitter located atop a high mountain or building. It receives signals on one frequency and relays or repeats them on a separate frequency in real time. By listening to the repeater's output frequency, one hears all stations using the repeater in a "one person transmits at a time" manner. Repeaters are financed and maintained by amateur radio clubs and operators using the repeater. An annual donation of $10 is usually more than sufficient to maintain repeater operating costs. Directories listing frequencies of FM repeaters throughout the nation are available from amateur radio dealers nationwide. Not all FM activities are via repeaters. Many amateurs also communicate on "simplex" frequencies set aside for "direct" chats. During normal times, repeaters are used for all types of casual conversations. Many repeaters also have autopatches, a feature that allows you to make non-commercial/non-business related telephone calls right from an FM handheld or mobile transceiver. A keypad on the radio is used for dialing the telephone, and special buttons such as "*" and "#" are used for turning the autopatch/telephone connection on/off. The sheer beauty of having a small FM transceiver that allows you to talk with other amateurs, monitor repeaters during inclement weather, call for help when needed, and make telephone calls on the spot is simply unequaled! Further, most newer handheld FM transceivers include extended receiver coverage for scanning/monitoring aircraft bands, public services (i.e., the police, fire, ambulance, etc.), and NOAA weather bands. Truly, they are a full communications center you can hold in your hand.

During inclement weather, FM repeaters often shift into "emergency net mode," and amateurs in all areas relay reports via the repeater. Amateurs with similar setups also join forces with local area weather bureaus, Civil Defense, Red Cross, etc. to almost instantly produce a complete network of information gathering and dispersal. Indeed, monitoring an area's 2-meter repeaters during emergency situations usually proves more fruitful than listening on a scanner or watching television reports. For survivalists living independently or traveling, VHF/UHF FM amateur radio is a priceless asset. Speaking of costs, incidentally, a deluxe 2-meter FM handheld is usually in the $200 to $350 price bracket. Similar FM mobile transceivers in the 50-watt output range usually cost between $300 and $500. Mobile antennas cost between $25 and $70. For home/base station operation, a mobile transceiver can be used with a power supply costing approximately $100. Home/base antennas range in price from a few dollars for a homemade version to $60 or $100 for elaborate commercially made versions.

Additional VHF/UHF activities are conducted on single sideband and usually involve a multi-mode transceiver. Using such radios, amateurs communicate by satellites on a national and international basis. Satellites are built by amateur radio volunteers, coordinated through AM SAT (Amateur Radio Satellite Corporation), and launched "piggyback" with commercial satellites. Other VHF/UHF activities such as Packet radio round out amateur radio's VHF/UHF "high band" world. These computer based and digitized messages are relayed by "digipeaters" around the country and have grown to form a massive network accessible almost anywhere. All one needs is an FM handheld, a laptop computer, and interface unit to assemble a completely portable packet station. Bulletin boards and electronic mailboxes are included in the packet network. Several low orbiting and easily accessible amateur satellites include packet capabilities and are therefore "flying mailboxes" that store and deliver their packets (electronically) to addressees worldwide. Amateur radio's VHF/UHF world obviously has a lot going for it! Now refer to Figure 4-11 and let's briefly discuss amateur radio's VHF/UHF bands.

LICENSE CLASS	TEST ELEMENT	TYPE OF EXAMINATION
Novice Class	Element 2 Element 1A	30-Question Written Examination 5-Words-Per-Minute Code Test
Technician Class [1,2]	Element 2	55-Question Written Examination and 3A (In 2 parts-30 Element 2. 25 Element 3A) (No Morse code requirement)
Technician Plus Class [2]	Element 3 Element 1	A 25-Question Written Examination if a Novice. A 5-wpm Code Test if a Technician.
General Class	Element 3B Element IE	25-Question Written Examination 13-Words-Per-Minute Code Test
Advanced Class	Element 4	A 50-Question Written Examination (No additional Morse code requirement)
Extra Class	Element 4B Element 1C	40-Question Written Examination 20-Words-Per-Minute Code Test

[1] No-Code License
[2] Effective 2/14/91

Note: Written examinations should be taken in strict ascending order of difficulty all the way to Extra Class. You should not be administered Element 3A until you have passed Element 2, etc. The code tests may be taken in any order. You can take the 20-wpm code test first if you can pass it. You can enter as a Technician without code, and then gain Technician Plus CW privileges by passing the Element 1A code test. You can enter as a Novice with code, and then gain Technician Plus by passing the Element 3A theory examination.

Figure 4-12 *Various classes of Amateur Radio licenses and their related examinations.*

2 Meters Truly, the most popular VHF amateur band. This is the "heartbeat" of FM activity around the country and around the world. With the exception of Novice licensees, amateurs of all license classes may operate 2 meters. The range of 144.0 to 144.500 is reserved for SSB and CW activities. The remaining range is used for FM repeater activities. Range depends on one's location and/or use of repeaters, but can generally be stated as 20 miles maximum for "direct" (no repeater use) and up to 150 miles by repeater relay. Some of the more popular repeater frequencies that are usually active include 146.94 MHz, 146.88 MHz, 146.76 MHz, and 146.98 MHz. The nationally coordinated "direct" frequency is 146.52 MHz. Antennas for 2 meters are usually quite short and vary from 18 inches tall to approximately 3 feet tall.

70 cm The next most popular band for FM and other VHF/UHF activities. It is available to all license classes except Novices. SSB, CW and amateur TV activities are usually conducted in the 420- to 440-MHz range, and FM activities are usually in the 440- to 450-MHz range. The communications distance of 70 cm is almost identical to 2 meters, and this band is also quite busy with repeater activity. Conversations on 70 cm are often more specialized and technical in nature than 2 meters.

222 MHz/1.25 Meters Another FM band used by all classes of licensees including Novices. Slightly less activity than 2 meters and 70 cm. It has approximately the same range.

6 Meters Situated midway between HF and VHF, 6 meters is often called the "magic band" because it can take advantage of "skip" conditions during times of unique propagation. Six meters is open to all licensees except Novices. The range of 50.0 to 50.1 MHz is reserved for CW operation. SSB activities are conducted between 50.1 and 52.050 MHz. The range of 52.050 to 54 MHz is used for FM and repeater activities. Antennas for 6 meters are typically 4.5 feet in height if vertical or almost the same size as a television for "beam designs."

The upper bands of 902 MHz/33 cm and 1240 MHz/23 cm support a variety of FM, SSB, and CW activities, primarily for amateur radio experimentation. These bands are being used for some satellite work, microwave study, etc., and are a fascinating frontier for today's electronic pioneers.

Equipment for Listening on the Amateur Radio Bands

As exemplified by discussions in previous paragraphs, tuning in amateur radio activities can supply you with a sheer wealth of information during times of unusual circumstances. The prime considerations are using a good communications-grade receiver capable of SSB and CW operation on HF or a good scanner for FM on VHF/UHF. The HF receiver must have a BFO (Beat Frequency Oscillator) for receiving SSB and CW. A creditable amount of band spread is also desired for easy fine tuning, and the receiver should exhibit good selectivity to reject adjacent frequency interference. Finally, a long-wire or outdoor antenna is desirable for reception of weaker signals. These criteria can occasionally be satisfied with a good grade portable shortwave radio; but a professional communications receiver made by companies such as Kenwood, Yaesu and Icom is more desirable. Communications receivers usually offer more operating assets or features such as variable bandwidth, IF shift or passband tuning, notch filtering, noise blanking/reduction, etc. An amateur radio HF transceiver can also be used receive-only style for monitoring the HF bands, but I hesitate to recommend them. The temptation to "talk back" (transmit) is simply too enticing! Assuming you are seriously studying for an amateur radio license, however, making a one-time investment in an amateur transceiver rather than a shortwave receiver is logical. In that case, select a transceiver that includes full reception coverage of the shortwave spectrum from 500 kHz to 30 MHz. Such transceivers are quite common on the amateur radio market, and perform equally as well or better than many of the top-end shortwave receivers. Herein lies yet another fact for the budget conscious. By clever investment in a wide-coverage HF transceiver and VHF/UHF mobile or handheld radio, one gains access to the full radio spectrum from the broadcast band to the microwave range around 1.2 GHz.

Although a dipole-type antenna cut to the particular band of interest provides the best reception, it is not always the best choice for general shortwave monitoring. However, stringing up a dozen wire antennas for different bands may also be illogical. The solution is a mild compromise. Trap or Windom-type wire antennas are fairly good for covering most of the bands discussed. Alternately, a simple long-wire antenna approximately 65 feet or even 33 feet in cramped spaces works quite well for listening. Just remember, longer wires (e.g., 120 feet) are good for listening on the lower bands between 160 and 80 meters, and shorter wires (e.g., 67 feet) are good for listening on bands up to 16 meters. Even shorter wire antennas (e.g., 16 feet) are good for monitoring between 16 meters and 10 meters.

A scanner such as those used for monitoring police, fire, emergency bands, etc. usually does a very good job of monitoring

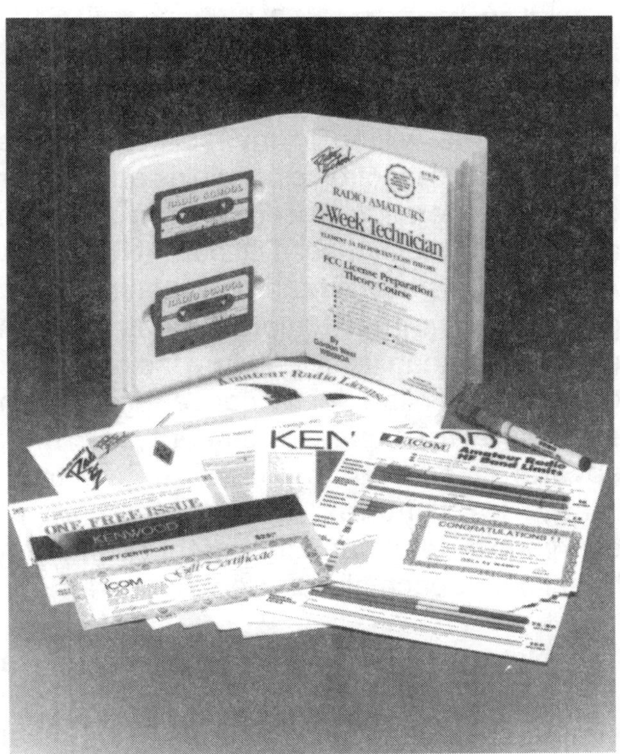

Figure 4-13 *One of the most well-known and proven study courses for amateur radio exams is produced by the Gordon West Radio School at 2414 College Dr., Costa Mesa, California 92626.*

amateur radio 2-meter and 70-cm repeaters. You simply load some of your area's repeaters' frequencies (found in a Repeater Guide from amateur radio stores) into the scanner's memories. Subsequent monitoring during emergencies can prove quite enlightening. It may, in fact, prove to be the final push you need to get a Code-Free Technician license as soon as possible! After obtaining an amateur license, you can purchase a transceiver with all the capabilities of the better scanners and also have two-way communications benefits. Scanners and amateur FM transceivers work well with basic antennas for general monitoring, but larger or outdoor antennas provide better access to fringe areas. Once again, a dazzling array of equipment and antennas is available from dealers nationwide.

Becoming a Licensed Amateur Radio Operator

Anyone willing to invest a few hours learning basic electronic principles and proper on-the-air operating procedures can obtain an amateur radio license. Difficult? Not really. Actually, the study is both enlightening and a good exercise in self-improvement. Why an exam rather than a flat fee? Anyone can plunk down a few dollars for a license; passing an exam is a true mark of sincerity. Further, several classes of licenses ranging from "easy" to "complex" are available to fit everyone's needs and interest. Let's take a brief look at each class. (See Figure 4-12)

Novice Class This introductory license consists of a 30-question written exam on simple theory and regulations. Technical electronic knowledge is not required. Inasmuch as limited access to HF bands is included in the privileges, a 5 w.p.m. code exam is also required. Operating privileges include voice/SSB in the popular 10-meter band, CW privileges within part of the 15-, 40- and 80-meter bands, and FM and repeater operations on the 220-MHz band. Power levels are restricted to 200 watts of output on HF bands.

Technician Class This exam consists of a 55-question test in two parts: one similar to the Novice exam and the other covering basic electronic theory. A Morse code exam is not necessary. Operating privileges include the use of the 6-meter/50 MHz and higher (VHF/UHF) bands. Technician class operators can enjoy 2 meter, 70 cm, and 220-MHz FM in addition to Packet, satellite, television, and SSB operating privileges.

Technician Plus Class In addition to passing the written examination for a Technician class license, passing a 5 w.p.m. code test allows one to upgrade to "Tech Plus." Privileges are the same as Technician and Novice combined. That is, the Tech Plus has access to all VHF/UHF FM activities and HF bands allocated for Novice use including 10-meter SSB for worldwide communications. If one is interested in only local area operations, the Technician license is fine. If worldwide communications is a goal, Tech Plus is the way to go.

General Class The "all-around standard license" for amateurs. This middle of the road license requires passing an examination on elementary electronic theory, on-the-air operating procedures, and a 13 w.p.m. code test. In return, you gain access to all amateur operating privileges across most parts of all HF bands and all VHF/UHF privileges. General class operators are also allowed full power output of 1500 watts on HF bands.

Advanced Class Requires passing an additional and somewhat more challenging written exam dealing with electronic theory. Advanced licensees gain access to additional frequencies within the HF amateur bands. An additional Morse code test is not required for this license; the General class 13 w.p.m. exam passed earlier is sufficient.

Extra Class This top-of-the-line license requires passing a written exam on advanced electronic theory and a code test of 20 w.p.m. The license authorizes full operating privileges on all frequencies on all amateur radio bands. A relatively small number of amateurs go all the way to Extra, possibly because they feel the additional privileges are not worth the effort. The Extra Class license truly stands as a tribute to the licensee. Please refer back to Figures 4-9 and 4-11 for clarification on frequency allocation privileges of each license.

There are three convenient ways of preparing for an amateur radio exam: 1. Attending classes conducted by radio clubs or radio schools; 2. Getting a complete package/course from an organization like A.R.R.L., W5YI or Gordon West Radio School; or 3. Self-study at home using license manuals available from dealers nationwide. Radio clubs around the country hold license classes on an almost continuous basis. A radio club may be listed in your area's telephone directory. Otherwise, check with the nearest amateur radio dealer for information on your area's clubs. Possibly, you may hear of a club by monitoring local FM repeaters on your scanner. Alternately, you might contact the American Radio Relay League's New Ham Hotline at 1-800-326-3942 for information on study guides and local amateur radio clubs in your area. The A.R.R.L. manuals and materials are also available from amateur radio dealers nationwide. Attending classes is the optimum way to prepare for an amateur radio license, but fitting it into a hectic lifestyle is often difficult. In that case, readers might check with the local radio club concerning weekend "quick courses" occasionally available in various cities. Next best is a complete home-study course as it puts all the necessary materials right at your fingertips and usually includes a few "extras" to encourage you along the way. Home-study courses also have the advantage of fitting in with impromptu lifestyles. They are especially good if you have a friend or spouse studying with you as a partner. In 1985, the F.C.C. set up a plan whereby a group of licensed amateurs could become F.C.C.-authorized Volunteer Examiners (VEs). The examiners are coordinated into 16 groups located throughout the country (See Figure 4-14). The only stipulation is that examiners cannot be involved in the "retail end" of amateur radio product marketing. Contact a Volunteer Examiner Coordinator (VEC) to learn of volunteer examiners in your area. We wish you good luck on passing the test and becoming a licensed amateur radio operator.

VOLUNTEER EXAMINER COORDINATORS

There are 16 volunteer examiner coordinator (VEC) groups authorized by the Federal Communications Commission (FCC), although some of these groups are small and offer only infrequent examination opportunities. The two main VEC groups operating nationally are the ARRL/VEC and the W5YI-VEC, included in the listing, below. Note the multiple day (D) and night (N) telephone numbers listed for some VECs; many of them are individual volunteer members' home telephone numbers.

American Radio Relay League and VEC
225 Main Street
Newington, CT 06111
Tel. 860-594-0300

Anchorage Amateur Radio Club
2628 Turnagain Parkway
Anchorage, AK 99517
Tel. 907-786-8121 (D),
907-243-2221 (N),
907-276-5121, or
907-274-5546

Central Alabama VEC, Inc.
1215 Dale Dr. S.E.
Huntsville, AL 35801
Tel. 205-536-3904

Golden Empire Amateur Radio Society
P.O. Box 508
Chico, CA 95927

Greater Los Angeles Amateur Radio Group
9737 Noble Ave.
Sepulveda, CA 91343
Tel. 818-892-2068 or
805-822-7473

Jefferson Amateur Radio Club
P.O. Box 24368
New Orleans, LA 70184-4368
Tel. 504-737-2315

Koolau Amateur Radio Club
45-529 Nakuluai St.
Kaneohe, HI 96744
Tel. 808-235-4132

Laurel Amateur Radio Club, Inc.
PO Box 3039
Laurel, MD 20709-0039
Tel 301-572-5124 (D),
301-317-7819 (N), or 301-588-3924

The Milwaukee Radio Amateurs Club, Inc.
P.O. Box 25707
Milwaukee, WI 53225
Tel. 414-466-4267

Mountain Amateur Radio Club
PO Box 10
Burlington, WV 26710
Tel. 304-289-3576 or
301-724-0674

PHD Amateur Radio Association, Inc.
P.O. Box fl
Liberty, MO 64068-0011
Tel. 816-781-7313

Sandarc-VEC
PO Box 2446
La Mesa, CA 91943-2446
Tel. 619-465-3926

Sunnyvale VEC Amateur Radio Club
PO Box 60307
Sunnyvale, CA 94088-0307
Tel. 408-255-9000

Triad Emergency Amateur Radio Club
3504 Stonehurst Pl.
High Point, NC 27265
Tel. 910-841-7576

Western Carolinas Amateur Radio Society VEC
5833 Clinton Hwy.
Suite 203
Knoxville, TN 37912-2500
Tel. 615-688-7771

W5YI-VEC
P.O. Box 565101
Dallas, TX 75356-5101
Tel. 817-461-6443

Figure 4-14 *The 16 amateur Volunteer Examiner Coordinators located in various areas of the United States. Contact a coordinator near you to determine when Amateur Radio license exams are administered in your area.*

Overview of Amateur Radio Equipment

The amateur radio market is quite extensive, and describing the full picture would require a separate book equal to (or larger than) this one. If you are presently a radio amateur and have thumbed through some recent ham catalogs, you are aware of that fact. If you are not a radio amateur, a brief glimpse of selections in HF transceivers and VHF/UHF handheld units will provide good insight into the wide varieties available to fit every need and lifestyle. I will therefore spotlight one unit from various manufacturers lines as I did in Chapter 3 (shortwave receivers), while progressing from simple to advanced and covering both HF and VHF/UHF transceivers. Unless otherwise noted, each manufacturer produces a full line of transceivers, handhelds, and accessories.

Yaesu FT-600 Illustrating the fact an amateur radio transceiver need not be large or complex to span the globe is Yaesu's Model FT-600 shown in Figure 4-15. This unit operates SSB and AM/voice modes and CW/Morse code mode on all 9 HF ham bands; plus, it makes a very good shortwave receiver with full coverage from 50 kHz to 30 MHz. The radio has 100 memories and can store often-used ham frequencies along with weather and international broadcast station frequencies in adjacent spots for instant access. The FT-600 is especially designed for switch-on-and-operate convenience. Front panel controls consist of volume, squelch, main tuning, and clarifier/RIT. The keypad on the front panel can be used for entering frequencies directly; and it has a secondary function of activating other features such as a noise blanker, selecting memories, etc.

Figure 4-15 Yaesu's Model FT-600 is built to rugged mil-spec standards. It is simple and easy to operate, and it makes a good transceiver for getting started in amateur radio. The unit operates all 9 HF bands plus receives all shortwave frequencies from 50 kHz to 30 MHz.

The FT-600 transmits a 100-watt signal, which is 10 to 20 times stronger than a typical CB radio, and it is capable of worldwide coverage. The transceiver operates directly from any 12-volt/20-ampere DC source, so it is equally suited for mobile or portable operation. The FT-600 may also be connected to its optional 110-volt AC power supply for base station/home use. For easy no-fumble operation in any environment and good survival preparedness, Yaesu's FT-600 fills the bill in fine style. Other models of deluxe, multi-featured transceivers are also included in the Yaesu line.

Alinco DX-70T This unit is a good example of the latest generation of ultra compact and full performance HF ham transceivers (Figure 4-16). The DX-70 is no larger than many CB radios, yet it sends out a 100-watt signal on SSB, AM, FM or CW, and covers all 9 HF bands. This unit also tunes in all frequencies from .5 to 30 MHz, providing double-duty service as an outstanding shortwave receiver. Additionally, the DX-70T transceives on the upper HF band of 6 meters, which is authorized for voice operation by Technician Class licensees.

Some of this go-anywhere transceiver's special features include IF shift for reducing interference, noise blanker, narrowband CW filter, 100 memories, preamp for boosting weak signals, and a very sensitive receiver.

Figure 4-16 *The Alinco DX-70T operates all nine HF bands with a stout 100-watt signal. It also receives all shortwave frequencies from 50 kHz to 30 MHz and has a detachable front panel for custom installation in a vehicle.*

Its additional features would easily fill this page. The DX-70T requires 13-volts DC at 20 amperes for transmitting at full output, or 13 volts at less current for reduced output or simply listening. The unit may be connected to an external power supply for home/base station use or powered directly from an auto's storage battery for mobile or portable operation. The front panel is removable and may be attached to an auto's dash with double-sided tape, while the main/RF section can be mounted under a seat or in the trunk for security. Extension cables for such custom installations are available in 15 and 30 foot lengths. Obviously, this unique transceiver exhibits a large amount of communications power in a small package. The DX-70T is presently Alinco's only HF transceiver.

Kenwood TS-870S In the category of full-size transceivers for home/base station use is Kenwood's Model TS-870S shown in Figure 4-17. Like previously highlighted units, this radio also covers all 9 HF bands and includes full shortwave coverage from 100 kHz through 30 MHz. The receiver section is exceptionally sensitive, the transmitter produces a solid 100-watt signal on all modes (SSB, AM, FM, CW and RTTY), and 100 memories store any frequency and mode. The TS-870S is loaded with more exotic features than I could possibly describe in this space, so I will highlight only some of the more interesting ones.

First, the unit's digital signal processor can automatically seek and notch out tones or "howls" interfering with the reception of a desired signal. Second, the D.S.P. can compare incoming information and determine what is noise and what are signals and literally pull stations out of background hiss. A large number of filter selections and adjustable filter shapes further enhance receiver capability; and a speech compressor adds additional "talk power" to the radio's 100-watt output signal. Other features include an automatic antenna tuner, noise blanker, electronic keyer for Morse operation, and multiple scan modes. The TS-870S requires a 13-volt/20-ampere source of energy for operation, and a mating power supply filling those requirements is optional. Truly a deluxe "rig"! Several other models of HF transceivers (most stepping down from the advanced-featured TS-870S) complete the Kenwood line.

Icom IC-781 This is a more upper end, all-mode HF transceiver covering all 9 bands plus receiving from 100 kHz to 30 MHz (Figure 4-18). Most of its deluxe features operate at the IF rather than AF level, which may be interpreted as "better" or "not quite equal to" units with DSP. In addition to 99 memories,

Figure 4-17 *This full-size Kenwood TS-870S is loaded with special features for HF communications and provides globe-spanning range with no problems.*

Figure 4-18 A deluxe do-everything HF transceiver, Model IC-781 from Icom, is especially designed for reliable communications amidst any type of band conditions.

speech processor, 150 watts output and automatic antenna tuner, the IC-781 can receive two stations on different frequencies simultaneously. A front panel "balance" control permits independent adjustment of the volume level of each station/signal. Rather than a dial or digital readout, the middle CRT display shows frequencies, modes, filter selections, etc. A lower portion of the display can be assigned for use as a spectrum analyzer indicating a received signal in the middle and signals up or down the band to the left or right. The display can also be switched to read out Morse code or RTTY text along the bottom, using an external decoder. An IF level notch filter and twin passband tuning plus numerous IF filters are some of this unit's additional features. The IC-781 operates directly from a 120-volt AC or 240-volt AC source. Additional models down from the IC-781 include mid and ultra-compact transceivers which round out the Icom line.

Alinco DR-150T, 2-meter FM Mobile As discussed earlier in this chapter, amateur radio's introductory Technician Class License provides access to the popular VHF and UHF bands such as 2 meters and 70 CM. These ranges are hot with continuous activities, and an amazing array of transceivers are available for them. Selecting a middle of the line unit for discussion, Alinco's Model DR-150T, 50-watt 2-meter FM transceiver, is shown in Figure 4-19. In addition to transmitting and receiving from 144 to 148 MHz, the DR-150T includes extended reception from 118 to 174 MHz plus 440-460 MHz (70 cm). It doubles as a scanner for aircraft, police, fire, ambulance, NOAA weather, and more. This mobile transceiver operates from a 13-volt DC 10-ampere source, has 100 memories, several scan modes, a private paging function, and a touchtone® keypad-equipped microphone for telephone autopatching. The unit's display also includes a spectrum scope that shows activity on a selected frequency in the middle and activity on adjacent frequencies (or memories) on each side. Other Alinco VHF/UHF models include basic 2-meter and 70-cm transceivers, 6-meter transceivers, and dual-band 2-meter/70-cm transceivers.

Standard C5900DA, 2-meter/70 cm/6-meter Mobile Shifting toward the upper end of FM mobile units, Standard Radio's triband C5900DA is shown in Figure 4-20. This transceiver not only operates the increasingly popular 6-meter band, but has extended reception of air, police and weather bands. Output power is 45 watts. The unit's list of

Figure 4-19 The Alinco DR-150T, 2-meter FM mobile transceiver delivers a 50-watt signal, receives aircraft, public service and weather bands, scans and includes a keypad on the mike for telephone auto-patching.

Figure 4-20 This triband Standard C5900DA FM transceiver operates 2 meters, 70 centimeters, and 6 meters with features galore. It also serves as a private mobile crossband repeater with remote control.

features is quite extensive, and I will highlight some in the "exotic" category. First, the front panel and main/RF sections may be separated and remotely mounted (as described with Alinco's DX-70T) for security. The C5900DA has 80 memories and may be expanded to 200 memories with an optional module. The C5900DA also operates full duplex, transmitting on one band while simultaneously receiving on one of the other two bands. Additionally, it has a crossband repeat mode which can receive. A low power signal from an owner's handy talkie on one band can be relayed at high power to a distant repeater on another band. Using the C5900DA in such an arrangement allows communication over a surprisingly long distance with a shirt-pocket-size talkie. Other features such as remote control, auto time out, and shutoff functions add operating security to a most elaborate transceiver. Other model mobile and handheld FM units "round out" Standard's line.

Kenwood TH-22AT, 2-meter FM Handheld Palm-size FM talkies such as Kenwood's Model TH-22AT shown in Figure 4-21 could easily be described as vital units for survival-conscious amateurs and individuals because they provide instant, on-the-spot communications. The TH-22AT transmits and receives on all frequencies in the 2-meter band, and its extended reception covers police, fire, public utility and weather bands in the 136- to 174-MHz range. Power output is 3 watts with its supplied rechargeable battery pack and may be increased to 5 watts with an optional 12-volt battery pack. Transceiver features include 40 memories, several scan modes, tone alert, built-in keypad for telephone autopatching, battery saver mode, and auto shutoff for forgetful owners. The handheld is especially designed for easy operation and may be personalized with accessories such as vinyl carrying cases, different size battery packs, rapid charger, and lapel speaker mike. Kenwood also produces several more models of VHF/UHF and dual-band FM handhelds and mobile transceivers.

Figure 4-21 The palm-size Kenwood TH-22AT, 2-meter talkie units like this are vital for personal communications in many survival situations.

Figure 4-22 The incredibly small Standard C508A dual band FM handheld has wide coverage and is closely akin to a full communications center you can carry in a shirt pocket.

Standard C508A Several amateur radio manufacturers produce palm-size and deluxe featured 2-meter/70 cm dual-band FM handhelds; however, I would like to show you a very interesting ultra-compact dual-bander. Standard Radio's C508A shown in Figure 4-22 measures only 3" x 2"' x 0.75" (H,W,D), operates 2 meters and 70 cm, and also receives 100-170 MHz, 300-480 MHz, and 900-950 MHz. The tiny transceiver also has 60 memories, several scan modes, auto shutoff, and operates with two regular AA cells. Considering the C508A's ability to tune in aircraft, marine, police, fire, ambulance, military, and weather services, in addition to operating on the hottest VHF/UHF amateur bands, it could easily be called a full communications center in your pocket. Standard Radio also makes additional models of VHF and UHF FM transceivers.

Hopefully, the previous overview gave you a quick insight into amateur radio transceivers. Now let's discuss some popular antennas for amateur radio.

Amateur Radio Antennas

An immense selection of commercially made antennas are available from amateur dealers nationwide; and they are ideal for individuals desiring plug-and-play operating convenience. Among the more popular types are verticals such as Cushcraft's 27 foot tall R-7000. This trim and unobtrusive antenna can be installed almost anywhere and does not require an external ground system. The R-7000 covers the 40-, 30-, 20-, 17-, 15-, 12- and 10-meter bands, and exhibits an approximate 3 dB gain, which means it makes a 100-watt signal sound like a 200-watt signal. A smaller version, Cushcraft's Model R-5, covers the 20-, 17-, 15-, 12- and 10-meter bands. It is only 17 feet tall and easily assembled or disassembled for quick in-the-field use.

For high performance base station use, beam antennas such as Cushcraft's A3S shown in Figure 4-23 are all time favorites. This is a directional type antenna covering the worldwide bands of 20, 15, and 10 meters; and it more than quadruples a transmitter's output power. A heavy duty support mast or a tower is required to support a beam like the A3S, and a strong antenna rotor is used for turning it. Now let's discuss some easy-to-assemble antennas all amateurs can make at home.

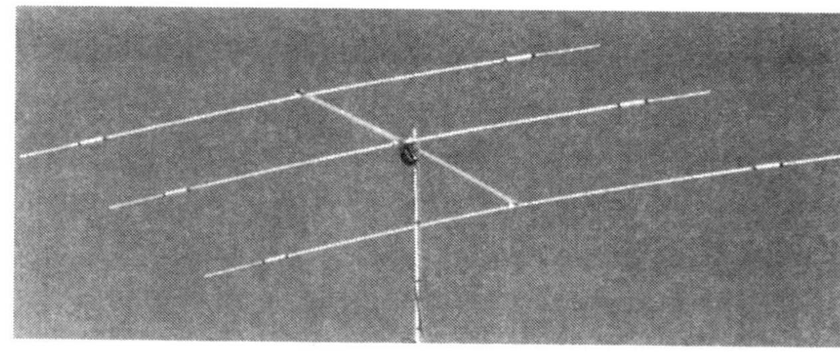

Figure 4-23 Beam antennas like this Cushcraft A3S are ideal for pumping a strong signal into a targeted area while minimizing interference from other areas. Antenna is 27' wide by 14' long.

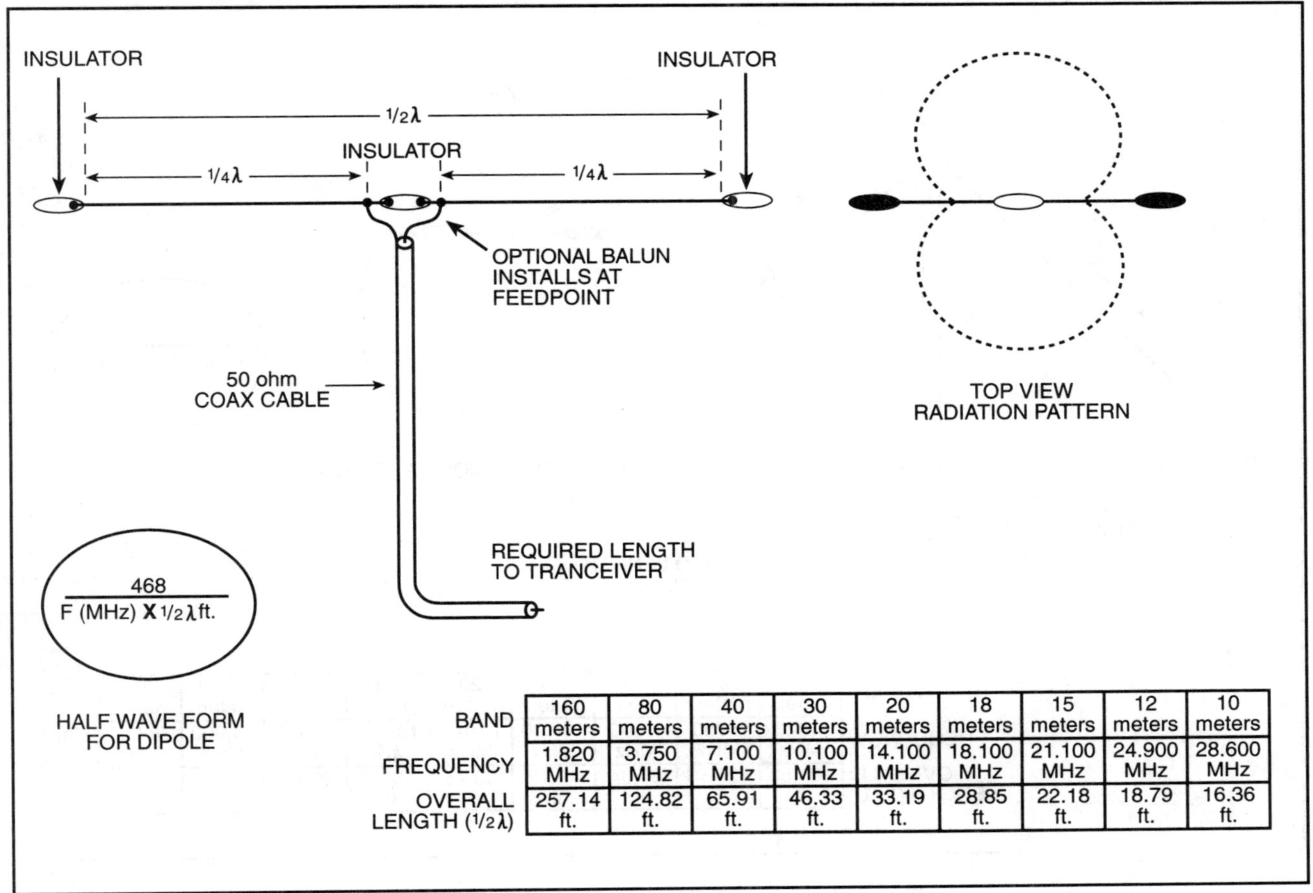

BAND	160 meters	80 meters	40 meters	30 meters	20 meters	18 meters	15 meters	12 meters	10 meters
FREQUENCY	1.820 MHz	3.750 MHz	7.100 MHz	10.100 MHz	14.100 MHz	18.100 MHz	21.100 MHz	24.900 MHz	28.600 MHz
OVERALL LENGTH (1/2λ)	257.14 ft.	124.82 ft.	65.91 ft.	46.33 ft.	33.19 ft.	28.85 ft.	22.18 ft.	18.79 ft.	16.36 ft.

Figure 4-24 *Dipole dimensions for HF bands.*

The Basic Dipole

The most common homemade antenna for the amateur HF bands is the basic dipole. It combines easy assembly with reasonably good performance. This antenna consists of a half wavelength of wire separated at its middle and is RF-fed by a two-conductor cable. Usual feed-point impedance varies between 50 and 75 ohms, thus establishing an acceptable match to 50-ohm cables such as RG58, RG8, or RG8X. Insulators are used at the middle of each end, and a non-conductive material such as nylon rope is suggested for suspension. If a popular commercial balun is used at the feed point, it can also serve as a center insulator. Baluns are quite useful for minimizing feedline radiation and ensuring close-to-theoretical radiation patterns. They are available from several amateur radio equipment manufacturers.

In order for the dipole to perform satisfactorily, it should be placed at least one-fourth wavelength above ground and in a reasonably clear location. This height is not difficult to achieve for operation on frequencies above 14 MHz, but it may prove slightly challenging when hanging antennas for 40, 80, or 160 meters. As a consequence of low antenna positions, ground losses and/or high radiation angles may restrict some of the potentially useful RF energy. The obvious solution for individual situations is to simply mount antennas in the clear and as high as feasibly possible; then accept unavoidable limitations gracefully and enjoy the results.

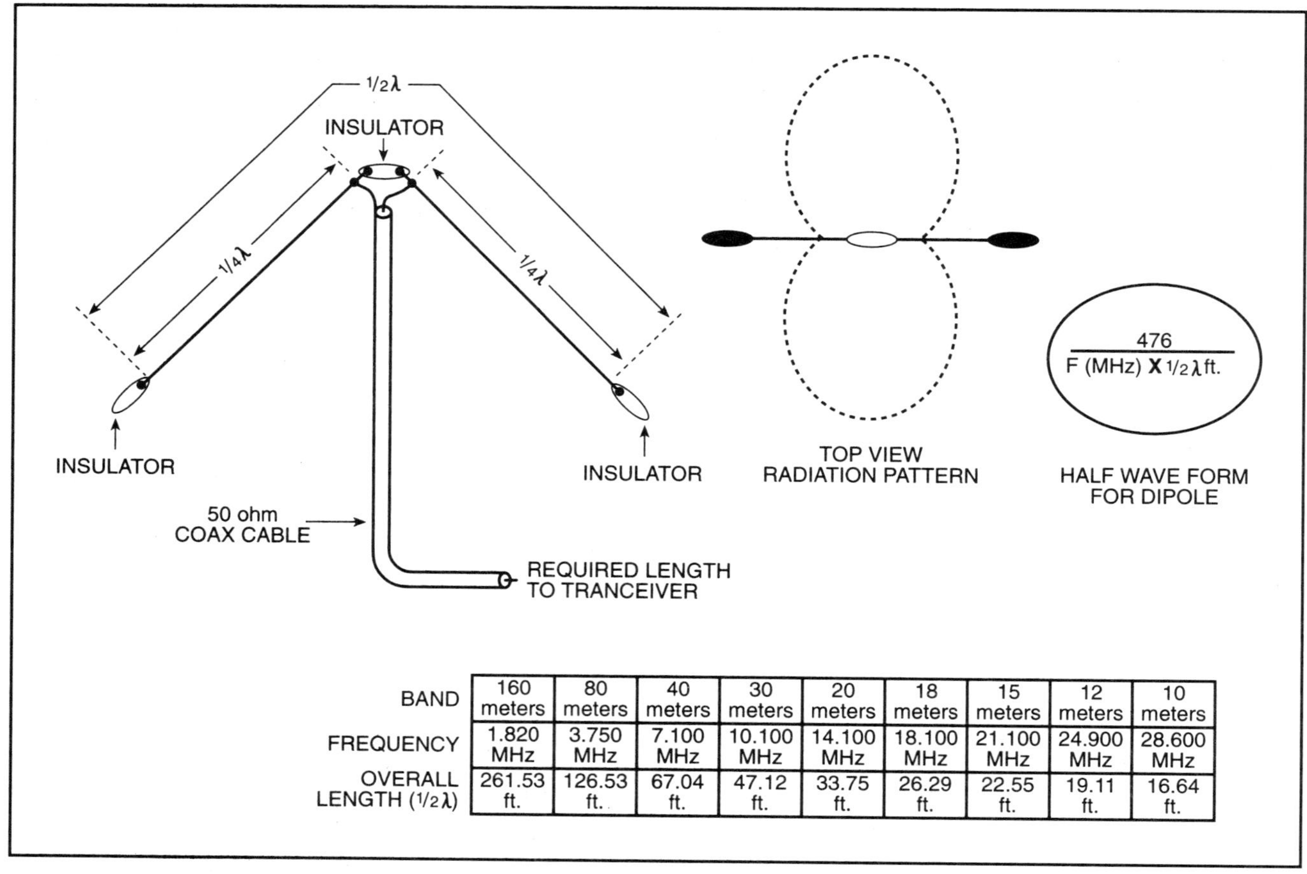

BAND	160 meters	80 meters	40 meters	30 meters	20 meters	18 meters	15 meters	12 meters	10 meters
FREQUENCY	1.820 MHz	3.750 MHz	7.100 MHz	10.100 MHz	14.100 MHz	18.100 MHz	21.100 MHz	24.900 MHz	28.600 MHz
OVERALL LENGTH (1/2λ)	261.53 ft.	126.53 ft.	67.04 ft.	47.12 ft.	33.75 ft.	26.29 ft.	22.55 ft.	19.11 ft.	16.64 ft.

Figure 4-25 *Inverted V dimensions for HF bands.*

Dipole dimensions for popular HF amateur bands are shown in Figure 4-24. These values may vary one or two percent depending upon your particular antenna's height and location. Realizing that such variations may be experienced with almost any antenna, we suggest measuring SWR, and pruning the final length as required. After pruning the dipole's length and confirming its operation within the desired range, seal all exposed connections with Coax Seal® to ensure long and reliable performance. Note: This weather protection step is very important and applies to all antennas. Dissimilar metals or wires joined together and exposed to moisture deteriorate rapidly. Likewise, exposed areas of coaxial cable absorb water and capillary action quickly spreads that moisture along the cable's length. Using tightly drawn electrical tape over the previously mentioned areas provides only temporary sealing.

The Inverted Vee

The inverted vee is a slightly modified version of the classic dipole, and it is a rather impressive DX performer for limited space and support situations. Each of the radiating wires is angled downward, providing low-on-the-horizon signal radiation while permitting installation on a single center support. Because the antenna's ends are near the ground, final frequency pruning is also easy. A high voltage peak is present at the antenna's ends when transmitting, so position them above the reach of small children. The angled elements also affect their resonant frequency; therefore, slightly longer dimensions than those for dipoles are used with the inverted vee. Inverted vee dimensions for popular HF amateur bands are included in Figure 4-25. Again, these values are subject to one or two percent variations with individual installations and require final pruning with an SWR meter for best results. A center balun is highly recommended for this antenna, and all exposed connections should be protected from the weather. In several ways, the inverted vee resembles a pair of leaning verticals, RF-fed at their tops. Similarly, the inverted vee performs best when erected in a clear "horizontal view" of at least one-half wavelength in all directions. Numerous variations of this basic design are possible and open to your ingenuity.

Twin-Lead Marconi

The twin-lead Marconi is an often overlooked but rather attractive low-cost radiator. Indeed, one well-known antenna manufacturer previously produced a multi-band/trap version of the twin-lead Marconi which was used and enjoyed by a number of radio amateurs. Dimensions and layout for the twin-lead Marconi antenna are included in Figure 4-26. The radiator is made of 300-ohm

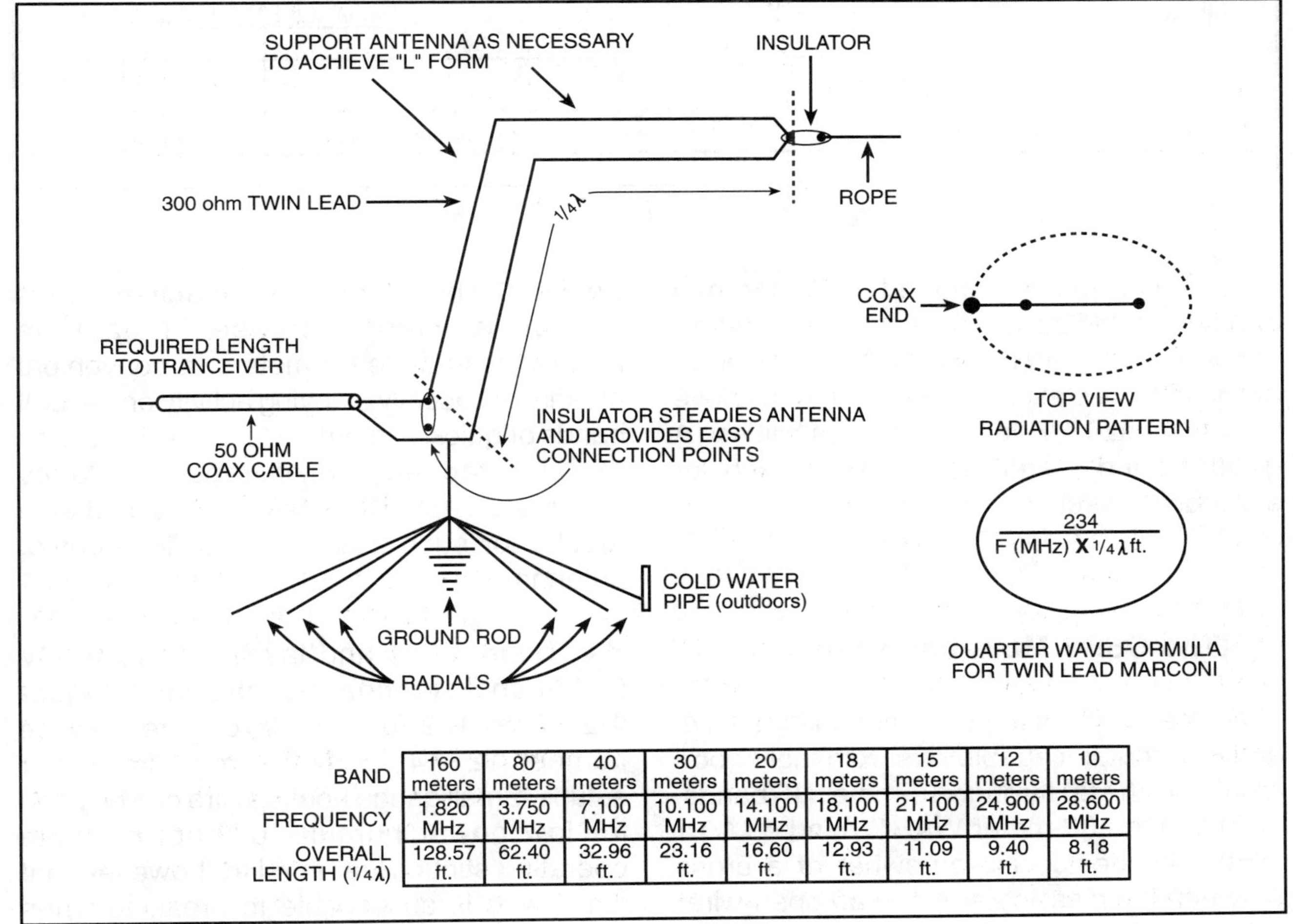

BAND	160 meters	80 meters	40 meters	30 meters	20 meters	18 meters	15 meters	12 meters	10 meters
FREQUENCY	1.820 MHz	3.750 MHz	7.100 MHz	10.100 MHz	14.100 MHz	18.100 MHz	21.100 MHz	24.900 MHz	28.600 MHz
OVERALL LENGTH (1/4λ)	128.57 ft.	62.40 ft.	32.96 ft.	23.16 ft.	16.60 ft.	12.93 ft.	11.09 ft.	9.40 ft.	8.18 ft.

Figure 4-26 *Twin-lead Marconi dimensions and layout for HF bands.*

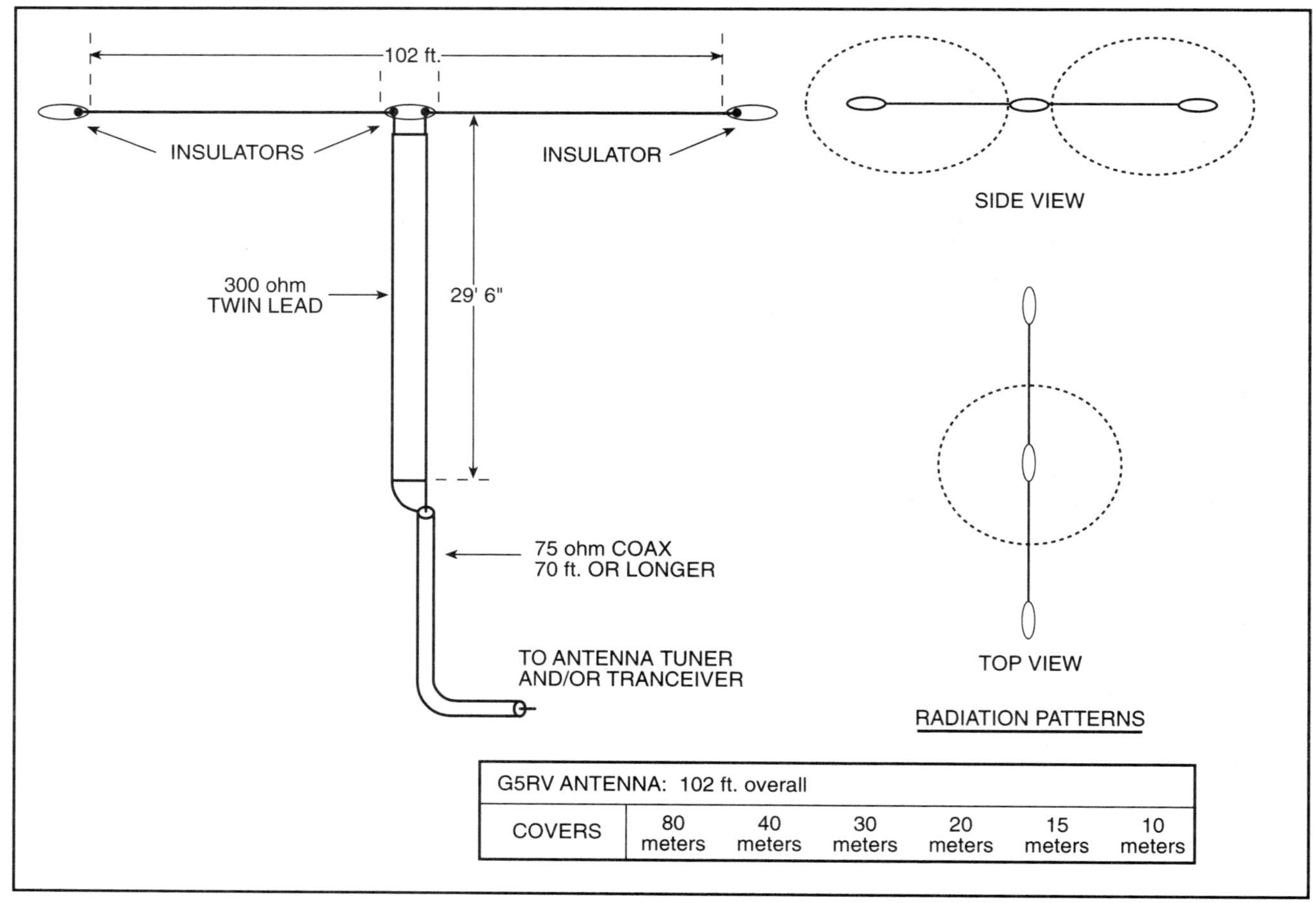

G5RV ANTENNA: 102 ft. overall						
COVERS	80 meters	40 meters	30 meters	20 meters	15 meters	10 meters

Figure 4-27 *The GRV multi-band antenna.*

TV twin lead which is shorted at the "far" end and connected between coaxial cable center and shield at the RF-fed end. Try to "run it" away from the station's gear rather than above it to minimize RF-feedback possibilities. A good ground system is the final touch for successful results.

The G5RV Multi-band Antenna

The G5RV is a unique and inexpensive antenna that is capable of surprisingly good results on the 80 through 10-meter HF bands, including the 30-meter WARC band on 10.1 MHz. The design is somewhat of a cross between the classic Zepp and an open-wire-fed doublet, and it is the creation of G5RV in England. This sky wire's attractive points include a slight increase in gain on frequencies above 40 meters, a convenient means for casually enjoying activity on several bands or quickly scanning for DX signals, and the sidestepped use of traps or a fancy transmatch. Actually, those are very favorable results for such a simple and "low-profile" antenna.

Dimensions and layout of the G5RV multi-band antenna are shown in Figure 4-27. The 102-foot length of wire may be number 12, 14, or 16 for medium power setups. Small gauge hookup wire can be used for low power pursuits or brief portable operating stints. Bear in mind, however, that small wire is susceptible to breaking when stressed between supports and affected by wind. When selecting 300-ohm twin lead, try to use the heavy-duty type with larger conductor wires (be sure, however, you do not choose the shielded type twin lead).

There are two minor stipulations and an "experimentally" flexible one regarding the G4RV multi-band antenna. Erect the antenna at least 34 feet above ground to allow the twin lead section to "fall" exactly vertically and end above ground level. Also, use 70 feet or longer lengths of coaxial cable between the twin lead and the station's transceiver to ensure smooth band-to-band tuning. Although 75-ohm coaxial cable is normally used with this antenna, a number of amateurs report switching to 50-ohm cable without any noticeable difference in performance. Either way, we suggest using an inexpensive antenna tuner (the "coaxial cable input/output" type is fine) between the transceiver and the antenna to provide a perfect 1:1 SWR on all frequencies. Otherwise, the SWR will probably rise to 2.5:1 or maybe 3:1 on some band edges (although the antenna will continue working properly). Finally, the G5RV's radiation pattern changes somewhat with the band of operation. This is no cause for alarm, the sky wire's overall radiation and DX capabilities are always good.

The Delta Loop Antenna

As most radio amateurs will attest, any antenna's performance in both transmitting and receiving is directly influenced by its

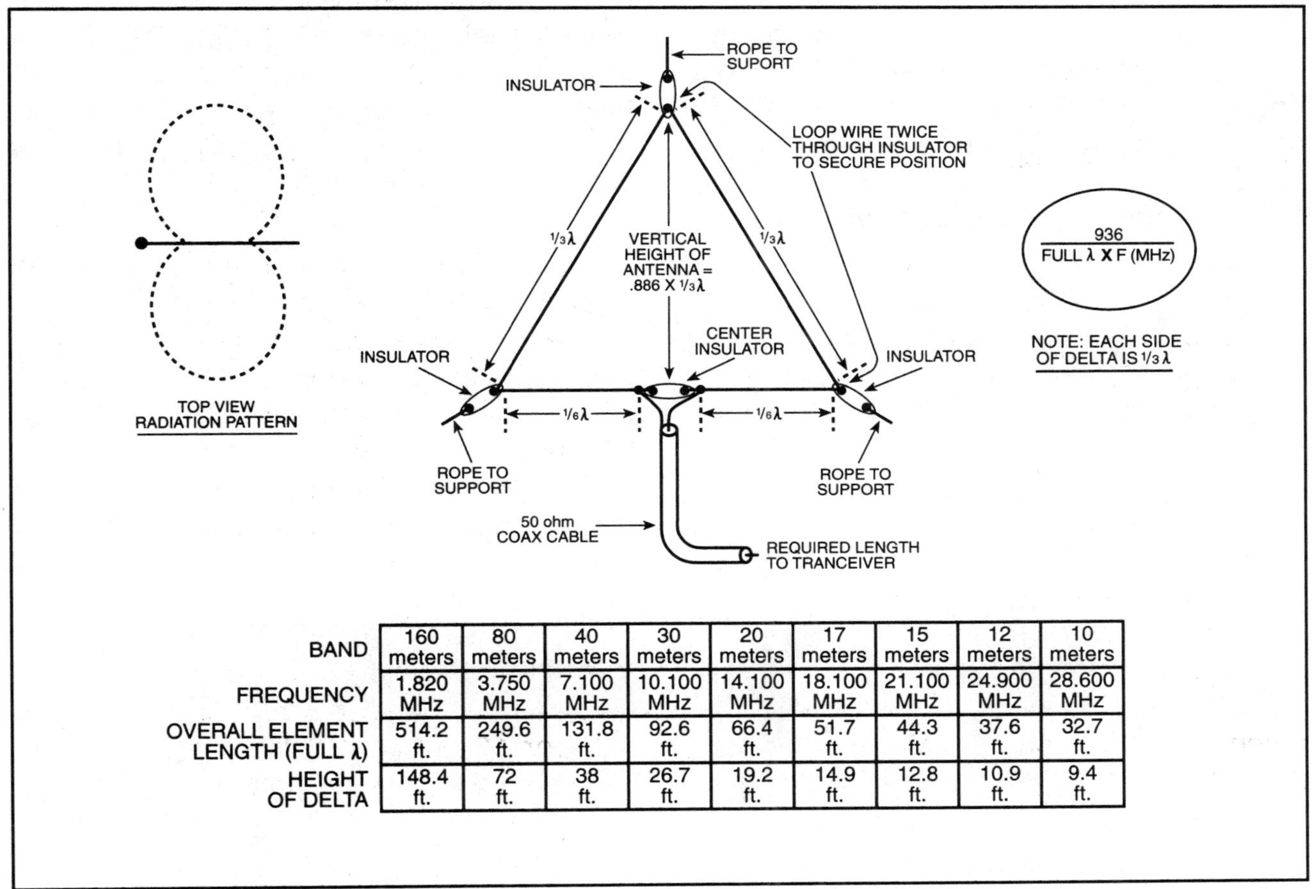

BAND	160 meters	80 meters	40 meters	30 meters	20 meters	17 meters	15 meters	12 meters	10 meters
FREQUENCY	1.820 MHz	3.750 MHz	7.100 MHz	10.100 MHz	14.100 MHz	18.100 MHz	21.100 MHz	24.900 MHz	28.600 MHz
OVERALL ELEMENT LENGTH (FULL λ)	514.2 ft.	249.6 ft.	131.8 ft.	92.6 ft.	66.4 ft.	51.7 ft.	44.3 ft.	37.6 ft.	32.7 ft.
HEIGHT OF DELTA	148.4 ft.	72 ft.	38 ft.	26.7 ft.	19.2 ft.	14.9 ft.	12.8 ft.	10.9 ft.	9.4 ft.

Figure 4-28 *Delta Loop antenna dimensions and layout for HF bands.*

overall capture area. Compacted antennas are compromises while oversized antennas are more desirable. Bearing this in mind, we present details of the Delta Loop antenna which is a full wavelength of wire configured in a triangle and RF-fed with 50-ohm coaxial cable. The attractive points of the Delta Loop are a low angle of radiation, good performance even when erected at rather low heights, large signal capture area, and installation which uses a single high support.

Layout and dimensions of the Delta Loop antenna are shown in Figure 4-28. The wire can be any conveniently available and reasonably strong material (twice normal length means twice normal weight). As a matter of convenience, we suggest initially "laying out" the antenna on the ground and forming its shape before erecting it. This will allow you to secure a top support/hanger bracket at the apex and one at each side (one end of an insulator works nicely—mark the spot, then tightly loop antenna wire through it twice and hold it securely in place). Next, fasten a long length of rope (previously drawn over your upper support) to the apex's insulator and raise the peak almost into position. You can then draw the corners outward with similar rope/insulator stabilizers and secure them to tie points near the ground. If the Delta Loop occupies sufficient free space, you may also be able to rotate the loop 90 degrees and take full advantage of its bidirectional pattern which is broadside to the loop. The Delta Loop's height can be calculated as follows: 0.866 times ⅓ full wavelength of wire. Note: This will be the antenna's maximum vertical size, not its height above ground. The antenna may be RF-fed at the center or at either side of the bottom "leg" as desired with negligible differences in performance.

In conclusion, we again emphasize amateur radio is an invaluable asset and invite (encourage!) everyone including (especially) survivalists of all interests to join the fascinating world of "ham radio." It can easily prove to be the best decision in preparedness communications you ever make!

CHAPTER FIVE

Scanning and Scanners

Survival communications are more vital today than ever before, and their roots actually transcend several decades. Since the early days of radio, people have been keenly attuned to monitoring activities of police, fire, ambulance, and other public services. Monitoring was also easier in the past, when important facilities used frequency allocations within the shortwave spectrum for their two-way communications. Anyone with a radio capable of tuning the broadcast band through 22 MHz could monitor their transmissions. Look at the dial on some radios from eras past (those calibrated in kilocycles and megacycles rather than kilohertz and megahertz), and you will see markings for international shortwave broadcast bands such as 49 and 31 meters. Look a little closer, and you will see markings for police, ships, aircraft, etc. Folks could listen to the news on one band, switch over to a shortwave band, and hear news as it was happening. The arrangement was ideal for survivalists, but criminals could just as easily listen to the police broadcasts in order to stay one jump ahead them.

A less accessible and more secure arrangement was obviously necessary, so police and other law enforcement agencies moved up in frequency to the then unused 150- to 160-MHz range. Soon thereafter, a well-known manufacturer, Hallicrafters, produced what we could call the first VHF monitoring receiver, the S-120. The desktop receiver tuned 118 to 180 MHz with a conventional analog dial, and it was quite popular among newspaper reporters and private detectives for keeping track of news and police activities. During that same time, aircraft started using the 118- to 136-MHz range, and a new near-shore maritime band opened in the 156- to 162-MHz range.

Figure 5-1 *High band scanners are available in a wide variety of models to fit every need and situation. Units are ideal for staying abreast of all local area activities.*

Numerous other facilities followed suit during subsequent years. As a result, the VHF range of 100 to 200 MHz began filling with a variety of activities. There are two noticeable differences between propagation of signals on HF and VHF. First, the 2- to 22-MHz range is good for long distance com-munication, but often lacks full coverage in local areas. Second, the VHF range provides more complete coverage of local areas and does not reach or skip into distant areas. As a result, VHF frequency/band allocations for different services can be "reused" approximately every 200 miles.

Figure 5-2 *Nearly all public transportation facilities are abuzz with VHF/UHF communications and security activities, and related action can be monitored directly on a modern scanner.*

VHF band scanners were the next evolution in monitoring. These units differed from earlier model radios by sporting a digital readout and keypad rather than a conventional knob and dial. The user punched in a frequency and pressed an Up or Down button for tuning. Bingo! The public at large gained access to all services in the VHF range. A vast number of organizations and agencies clamored to the newly pioneered range. Scanning became a national craze. Portable and handheld scanners also entered the picture.

Approximately a decade later, police and law enforcement agencies in larger cities migrated up to the then unused UHF range of 450 to 470 MHz. This UHF range was attractive because of its privacy aspect. Also, higher frequency UHF signals reflected and bounced more among tall city buildings and provided more reliable communications in metropolitan areas. Scanners covering both VHF and UHF were the next evolution. Soon, the cell telephone craze hit "full force." Scanner ranges were then extended to cover from 30 MHz to 1000 MHz/ 1 GHz. Next, Congress passed the Electronic Communications Privacy Act of 1986 prohibiting off-the-air monitoring of cell phone conversations. Today, new model scanners are required to "lock out" the cell phone range of 870 to 896 MHz. Naturally, a number of "unlocking" kits/modifications for radios to cover the cell phone range are now being marketed in a somewhat controversial manner.

In today's world, scanning is hotter than ever before. Indeed, it is a vital part of any survivalist's equipment for tuning in to what is happening both near and far from home base. Scanning puts everything local right at your fingertips. There is NOAA weather, police, fire, ambulance, and Amateur Radio communications available for monitoring almost continuously. Railroad com-munications can be monitored during transportation accidents and/ or chemical spills. ATF and border patrol plus U.S. Customs can also be monitored for drug smuggling and illegal immigration activities. Ships near shore, airports, aircraft, and much more are available for monitoring at the touch of a button. Migrations to still higher frequencies continue today. Most recent is the "800-MHz band explosion." Not only are voice

Figure 5-3 *Deluxe base-station scanners/receivers like Icom's R-8500 shown here are loaded with features and can even find all the action for you.*

communications in the 810 to 816 MHz range growing in number each day, police and other public safety agencies are using trunking systems in the 856- to 867-MHz range. Trunking is the first serious approach to privatizing and more efficiently routing special communications. Vehicles and base units in a trunking system are under direct computer control. A mobile or base unit initially transmits on any open frequency within the trunking system and a related computer adds addressing tones to the transmission. When another unit responds or replies, a separate frequency within the trunking system may be used while computer addressing ensures proper routing to the desired station. As you can surmise, modern-day scanning and monitoring is a genuine cat-and-mouse game. Technology is continuously improving to meet the challenge. Two of the most recent innovations, for example, are hyper (super fast) scanning and units with more than 100 memories for keeping track of everything—almost at once. Some of the very latest scanners even zip from HF through UHF in only a few seconds time and load all active frequencies into memories. Just set the unit on scanning, leave it for a period of time, then return and check memories according to "hits" (number of times active) to find action in any area or for any abnormal situation.

Scanning and the Law

Many types of communications fill the VHF/UHF spectrum and a full-range scanner is capable of unlimited reception. New scanning enthusiasts, however, should be aware of a few regulations and stipulations regarding monitoring. This is not intended to discourage or intimidate you, but simply to inform you of your legal and ethical responsibilities while listening.

First, you are considered a "third party" when listening to two-way conversations between others. Remember, the Communications Act of 1934 prohibits such third party listeners from revealing the contents of private two-way messages intercepted off-the-air. Additionally, a third party listener cannot use information in such two-way messages for personal gain. More specifically, a related excerpt from the Communications Act of 1934 reads as follows: *"No person, not being authorized by the sender, shall intercept and divulge or publish the existence, contents, substance, purport, effect, or meaning of such intercepted communications to any person. No person not being entitled thereto shall receive or assist in receiving any interstate or foreign communications by radio and use such communications for personal benefit or for the benefit of another not entitled thereto. No person having received any intercepted radio communication or having become acquainted with the contents, substance, purport, effect, or meaning of such communication knowing that such communication was intercepted, shall divulge or publish the existence, contents, substance, purport, effect, or meaning of such communication or use such communications for his own benefit or for the benefit of another not entitled thereto."*

In other words, monitoring is permitted provided you do not discuss what you hear with others or use acquired information for personal gain. Listening to police calls, then informing fugitives or news reporters is one example of illegal monitoring. Listening to a police call regarding a stolen auto used in a robbery, noting a matching auto with coinciding tag number nearby, then reporting this to police (but not "getting involved!") borders on being illegal. Such reports are usually acceptable, however, provided the reporting party is an asset rather than an hindrance to authorities.

Second, scanner owners should be aware of the Electronic Communications Privacy Act of 1986 restricting third party monitoring of cellular telephone frequencies. Some of the notable excerpts from ECPA 1986 follow.

"Interception and disclosure of wire, oral or electronic communications are prohibited. Except as otherwise specifically provided herein, it is unlawful for any person to:

A). Intentionally intercept, endeavor to intercept or procure any other person to intercept or endeavor to intercept any wire, oral or electronic communication.

B). Intentionally use, endeavor to use, or procure any other person to use or endeavor to use, or procure any other person to use any electronic, mechanical or other device to intercept any oral communication when

1). Such a device is affixed to, or otherwise transmits a signal through, a wire, cable, or other like connection used in wire communications or

2). Such a device transmits communications by radio, or interferes with the transmission of such communications or

3). Such person knows, or has reason to know, that such device has been sent through the mail or transported interstate or foreign commerce."

Stated another way, anyone caught listening to cellular telephone frequencies is breaking the law. ECPA 1986 actually did more to kindle interest in cell phone monitoring than it did to minimize it. Consequently, the cell telephone industry secured congressional passage of the Telephone Disclosure and Dispute Resolution Act in 1993. Specifically, it prohibits the production or importation of scanners or receivers capable of tuning or being modifiable to tune cell telephone frequencies after 1994.

Where does that leave us today? If you have a scanner made prior to 1994 or one capable of tuning cell telephone frequencies, you will probably do some "curious monitoring." Simply remember not to repeat intercepted conversations. The curiosity will wear off after a day or two of monitoring.

Figure 5-4 *An endless variety of books, magazines and frequency guides for scanning enthusiasts are available from shortwave radio dealers and bookstores nationwide. They are ideal information sources for determining where to find activities in various areas.*

Finally, bear in mind some states have their own private laws regarding scanners. New York and New Jersey, for example, deem mobile scanners capable of receiving police calls illegal. Florida prohibits "installation" (?) of a receiver capable of tuning police frequencies. Check with your state's laws, monitor wisely, and do not "brag" about what you hear. Simple enough?

Finding the Action

Considering the large number of services operating on VHF/UHF frequencies, a newcomer might visualize tuning for a specific activity as virtually impossible. Fortunately, such is not necessarily the case. Most services communicate within a preestablished range of frequencies delegated for their use. Also, scanning directories and magazines with up-to-date frequency guides are published on a regular basis. As additional guidance, a general range "getting started" guide is included in Figure 5-6 for your convenience. Program upper and lower frequency limits of a service you wish to monitor into your scanner, set it on high speed scan, and let it find the action for you. Keep in mind that all frequency ranges shown are approximate and variations are to be expected. Occasionally, scanning bands for unexpected activity also proves fruitful. As an example, some inland police services have been noted using marine band frequencies and marine radios in a legally licensed manner, and some off-duty officers on security patrols occasionally use GMRS equipment. Finally, remember patience is a virtue when scanning for activities like stakeouts. Additional details will be presented as we proceed. Let's now focus on the Frequency Chart.

Alcohol, Tobacco and Firearms (ATF) These VHF (165-168 MHz) and UHF (406-420 MHz) ranges are hot spots of activity during times of arms confiscations and various smuggling activities. Simultaneously, other frequencies used by the police and FBI are also active during ATF operations.

Figure 5-5 *The practice of monitoring cell telephone conversations was deemed illegal by the Electronic Communications Act of 1986. Modern VHF/UHF receivers and scanners thus lock out or skip coverage of cellular phone frequencies.*

Amateur Radio The 2-meter/VHF (144-148 MHz) band is ham radio's "hot spot" of activity today. It stays abuzz with repeater-oriented conversations almost continuously. Set your scanner to the 146-148 range for monitoring. This band proves ideal for monitoring during abnormal weather situations. The 1.25-meter (222-225 MHz) band is also active around the country on FM, and it supports substantial repeater activity. The 70-cm band (420-450 MHz) is almost as active as 2 meters, and it usually supports more specialized and technically oriented conversations. Most action occurs between 445 and 448 MHz. The upper

VHF/UHF COMMUNICATIONS THAT CAN BE MONITORED ON SCANNERS

Alcohol, Tobacco, Firearms
165-168 MHz (ATF)
406-420 MHz

Amateur Radio
144-148 MHz
222-225 MHz
420-450 MHz
902-928 MHz

Ambulance
155.0-155.5 MHz
462-468 MHz

Aircraft
118-128 MHz (Air Traffic Control)
123.5-123.7 (arrival and departure)
Note: Above aircraft use AM mode

Border Patrol
408-417 MHz

Business Bands
150-162 MHz
461-465 MHz
502-512 MHz
851-853 MHz

Cellular Phones
870-896 MHz
Note: Commercial aircraft cell phones use 894-896-MHz range

Cordless (home) Phones
46.6-49.9 MHz (older models)
900-902 MHz (newer models)

FBI
162-167 MHz

Federal Emergency Management Agency (FEMA)
166.2-170.2 MHz
Also 138.225, 141.725, 408.40, 410.48, 413.21, 417.66 and 417.05 MHz

Fire
154.0-154.5 MHz
451-454 MHz

General Mobile Radio Service (GMRS)
462.5-467.7 MHz

Military Aircraft
236.4-236.7 MHz (Air Traffic Control)
243.0 MHz (emergency)

National Oceanic and Atmospheric Administration (NOAA)
162-163 MHz

Naval Aircraft
277.5-278 MHz
340.2 MHz (Air Traffic Control)

News Media
450.0-455 MHz

Nuclear Regulatory Commission
167-172 MHz

Police
154-159 MHz (smaller cities and rural areas)
453-454 MHz and
460.0-460.6 (larger cities and metropolitan areas)
and 810-816 MHz (new 800-MHz band activity)

Police and Municipal Trunking
856-867 MHz

Railroads
159-161 MHz

Secret Service
164-167 MHz

Strategic Air Command (SAC)
311.0 MHz

U.S. Customs
165-166 MHz

Figure 5-6 *VHF/UHF communications that can be monitored on scanners.*

UHF band (902-928 MHz) is relatively new, and activity is growing on a daily basis. Presently, it is good for "curious monitoring."

Ambulance Both on-the-scene and en route communications can be heard within these ranges daily. When scanned in combination with police and fire frequencies, it can provide good insight to "what's happening on the scene." The VHF range of 155 MHz is very popular in smaller cities. It may also include lifesaver helicopters. The UHF range of 462-468 MHz is more common in larger cities and metropolitan areas.

Aircraft Control towers usually communicate with airplanes between 118 and 128 MHz. Aircraft arriving and departing communicate with the tower using frequencies between 123.5 and 123.7 MHz. En route aircraft communicate with various towers using the 300 to 360 MHz range. Note: aircraft use AM mode, whereas other VHF/UHF services use FM mode.

Border Patrol The range of 408-417 MHz is often used between Texas and Mexico. It becomes a hotbed of activity during times of illegal immigration and smuggling.

Business Bands These bands are used by large corporations and smaller businesses alike. They may include such services as heating and air conditioning companies, tree trimmers, delivery services, and construction companies. All ranges see almost identical activity, with the exception of 851-853 MHz. This newer band requires more state-of-the-art equipment, and companies continue using existing equipment as long as possible before upgrading. Surprisingly, some smaller city police services have been legally authorized to use business band frequencies for communications. Municipal utility companies also use various frequencies in business bands. Check the range of 150-155 and 461-463 for activity in your area.

Cellular Telephones This 870- to 896-MHz band is the "off limits" range that is illegal to monitor and is being "locked out" on recently manufactured scanners. We included it here so you will know what not to monitor! Have you noticed the cellular telephones in seat backs aboard commercial aircraft? You can telephone home using your credit card. These airborne cell phones operate like conventional cell phones, and transceive in the upper cell range of 894-896 MHz. Monitoring this range is also illegal.

Cordless (Home) Phones Wireless telephones are the most popular electronic convenience today. Due to their extremely low power, range is usually limited to 100-700 feet. Scanners in high locations or amidst large apartment complexes, however, often receive many cordless telephone conversations. The legality of monitoring these calls is a ticklish situation. Newer and more expensive cordless telephones usually operate in the 900-902 MHz range. The latest digital models include voice encryption to avoid unauthorized access and monitoring.

FBI The Federal Bureau of Investigation has field offices in almost every major city throughout the United States. Some communications are via voice/FM, some are scrambled, and some are via Morse code (CW). Activity is variable and shifts between frequencies in an unpredictable manner. Use your scanner's high-speed scan mode to spot the action, and maintain patience in monitoring. A tape recorder with voice activated operation (VOX) and extended recording time may prove useful.

FEMA The Federal Emergency Management Agency swings into operation during occurrences like the Oklahoma City bombing, airplane crashes, and other serious incidents. They use a wide variety of frequencies, and are usually difficult to locate. Our listing shows some of their more often utilized channels/frequencies.

FEMA's main purpose is management during crises and disasters. Consequently, they coordinate activities among as many as 26 federal agencies. These include police, fire, transportation, communications and engineering, medical services, search and rescue, and more. Being responsible for overall management in a wide range of environments, they use numerous com-munications media—many existing at the particular scene. Some FEMA teams carry up to a dozen handheld talkies. They may also carry a portable 1.5-GHz Immasat satellite transceiver for direct communication with FEMA headquarters in Washington, DC. FEMA also has its own fleet of multi-radio vans fully equipped with transceivers covering from the broadcast bands through 2 GHz continuously!

Fire Departments Frequencies within both VHF and UHF ranges are used by fire departments. Fire dispatchers usually operate on a separate frequency from crews actually on the scene. Further, handhelds used by firefighters are usually low power and cannot be received for more than three or four miles from the scene. Monitoring fire, police and ambulance services simultaneously during an emergency provides a good "electronic overview" of precise events.

GMRS The General Mobile Radio Service is used by various groups including security guards at banks and malls, construction crews lacking a business band license, and even families wishing to keep in touch with each other while vacationing. License requirements are minimal.

Police Generally speaking, most larger police departments use 453-460 MHz for routine communications and suburban/smaller departments use 154-159 MHz. Lists of frequencies for specific departments are often available from dealers of scanning equipment in their related cities. If you do not have a frequency guide, I suggest programming a police department's upper and lower limits into your scanner; then wait for it to find active channels to load into memories.

Once a particular precinct is found, other precincts usually fall within the same megahertz segment. Police departments typically use high-powered repeaters to ensure full coverage of their precinct/area, so their signals are usually quite strong. Initially monitoring during different hours of the day and evening usually proves beneficial in finding active police repeaters. Short bursts of tones at the beginning of each transmission have proven to be a good tip-off to police communications.

A unique concept known as trunking is being implemented by police and other municipal organizations in larger cities. Simply explained, it allows police and other related departments to work in conjunction within a specified area which would otherwise be limited by lack of inter-departmental communications. For the monitoring scanner, it means you may find a police call on one channel and a street sanitation department on the same channel two minutes later. Possibly a good technique for trunk monitoring of a specific service would involve using two or three scanners on fast scan—one scanning up from an active frequency, one scanning down, and the third one staying on the active frequency. In a trunking system, no single service is assigned a specific frequency. When a police officer transmits from his vehicle, a computer will select an unused frequency. When another public service, such as a street department, transmits with their mobile unit, a computer automatically selects the next unused frequency. A group of up to 30 frequencies may be involved in a trunking system.

Military Aircraft Military bases and training centers located throughout the country maintain their own control towers and communications systems operating in a discreet range of frequencies. Sharp monitors may hear training exercises, practice air battles, and more in the range of 236-243 MHz.

Naval Aircraft The Navy utilizes a separate band of frequencies for its aircraft and monitoring 277-340 MHz is suggested. You may hear anything from aircraft carrier communications to Blue Angels, provided you are in the approximate vicinity of the communications.

Strategic Air Command This elite aircraft group has been reported in the 311-MHz range and proves to be fascinating listening.

News Media Used by news reporters and vans, traffic reporters, television station "instacams" and many more. Good for monitoring on-the-scene reports with pocket scanners.

Nuclear Regulatory Commission Often used around nuclear power facilities. Also active in any area dealing with nuclear materials.

NOAA The National Oceanic Atmospheric Administration maintains a network of over 365 weather broadcast stations throughout the United States. Most are situated in the 162.550- and 162.450-MHz channels. During inclement weather, monitoring NOAA and Amateur Radio repeaters on 2 meters keeps one current on what is happening. An additional chapter in this book is dedicated to NOAA, and you are encouraged to read it for additional information.

Railroads Activity among various railroads can be monitored almost 24 hours a day. You may hear railroad police officers, inspectors, switching towers and much more. Both commuter and freight trains use channels within our listed 159- to 161-MHz range. If you are near any railroad of any size, you are aware of the importance of the railroads to our country. The railroads are a major mover of most forms of energy—fuel oils, petroleum and propane products, etc., and they move coal products that produce the electrical energy we use twenty-four hours a day. Along with the transportation of our energy sources, the railroads transport a major share of our basic food products from the farms to our cities for conversion and processing into finished food products.

The railroads transport large amounts of commodities used in most industries, in addition to finished products from our factories. In all walks of life, our railroads are still extremely important to our country. Our military will transport critical equipment and supplies via rail to seaports and military bases throughout the United States.

The railroads carry their communications on the VHF and UHF frequency bands. In addition to many other frequencies and communications systems, the railroads have over 90 assigned channels on VHF with frequencies between 160.215 and 161.565. The railroads have informative and interesting communications on VHF from their local freight yards and terminals to trains and from the dispatchers and control towers. Another interesting listening area is the railroad's police section which controls the security of the company's rails. When there is something unusual going on, this can be fascinating and inter-esting listening.

An excellent pub-lication devoted solely to railroads, The American Railroad Frequencies Book, can be obtained by writing to P.O. Box 1612, Waukesha, WI 53187.

Secret Service Secret Service offices are maintained in most major cities. Communications are via FM, CW (Morse code), and occasionally en-crypted voice.

U. S. Customs The range of 165-166 MHz is quite active around all borders and coastal areas of the United States. Related communications range from following smugglers and illegal immigrants to seizure of drugs, firearms, and more.

Maritime Mobile Large ships and small crafts navigating within a few miles of shorelines use maritime mobile FM transceivers operating within the 156- to 162-MHz range for localized communications. As illustrated in Figure 5-7, the range is divided into 88 channels. Most maritime transceivers are marked only with channel numbers rather than frequencies, so a pleasure boat operator may not realize/know the exact frequency of use. Notice in our maritime mobile list that full duplex (separate transmit and receive frequencies) operations are utilized on channels 78 through 88. Two scanners, one on each frequency, are required for hearing both sides of related conversations on these channels. Now some additional notes for your monitoring knowledge. Channel 16/ 156.80 MHz is the distress frequency for emergencies. Channels 12, 20, 65, 66, 73 and 74 are used for port operations. The Coast Guard uses channels 21, 22 and 23. The Coast Guard's auxiliary full duplex channel is 83. Pleasure boats use channels 68 through 72. Marine telephone calls, which are handled through a shore-based operator, are conducted on channels 24 through 28 and 84 through 87.

Channel	Frequency (in MHz)	Channel	Frequency (in MHz)
1	156.05	60	156.025 and 160.625
2	156.10 and 160.70	61	156.075 and 160.675
3	156.15 and 160.75	62	156.125 and 160.725
4	156.20 and 160.80	63	156.175 and 160.775
5	156.25 and 160.85	64	156.225 and 160.825
6	156.30	65	156.275 and 160.875
7	156.35	66	156.325 and 160.925
8	156.40	67	156.375
9	156.45	68	156.425
10	156.50	69	156.475
11	156.55	70	156.525
12	156.60	71	156.575
13	156.65	72	156.625
14	156.70	73	156.675
15	156.75	74	156.725
16	156.80	75	not assigned
17	156.85	76	not assigned
18	156.90	77	156.875
19	156.95	78	156.925 and 161.525
20	157.00 and 161.60	79	156.975 and 161.575
21	157.05	80	157.025 and 161.625
22	157.10	82	157.075 and 161.675
23	157.15	83	157.175 and 161.775
24	157.20 and 161.80	84	157.225 and 161.825
25	157.25 and 161.85	85	157.275 and 161.875
26	157.30 and 161.90	86	157.325 and 161.925
27	157.35 and 161.95	87	157.375 and 161.975
28	157.40 and 162.00	88	157.425 and 162.025

Figure 5-7 *The 88 VHF band maritime mobile channels.*

Our previous discussion covers dozens of services and hundreds of frequencies to give you an accurate overview of the unlimited potential of scanning/monitoring VHF/UHF bands. As you spend time monitoring and tuning in services of particular interest, recognizing "hot spots" of activity will become second nature. Really! Scanning is actually easier than it seems—and most informative.

How To Select Scanning Gear

An impressive array of equipment and accessories comprise today's scanner market, and newcomers may become overwhelmed with choices when making a buying decision. Hopefully, our following discussion will provide insight in that direction and simplify the process.

First, I suggest visualizing your monitoring needs and considering whether a basic and easy-to-use unit or a more complex "do everything" scanner best fits your lifestyle. Next, review technical specifications listed in a unit's ad, literature, or instruction manual to ensure it will fill your needs. Finally, consider whether a 120-volt home unit or a battery-powered scanner best serves your intended purposes. The latter consideration is probably the most important, as home AC power is usually not available when a scanner is most needed. Now let's take a closer look at each of the previous factors.

A good way to visualize your monitoring needs is by first making a list (including frequencies) of the services you wish to hear. If police, fire and weather in your immediate area are the only interests, a basic scanner covering that range of frequencies may be ideal. If you travel into unknown areas or live in a large metropolitan area, greater frequency coverage plus super-high-speed scanning and a large number of memories (over 100) will probably be desirable for following brief transmissions and multiple services "rapid fire" order. When purchasing a first scanner, some people feel dependent on a salesperson's guidance. If the salesperson really knows his/her business and is possibly also a scanning enthusiast, the advice and suggestions given may be useful; however, such knowledgeable salespeople are a rare breed. Ask the salesperson to punch in some of your selected frequencies or, better yet, try punching in a few frequencies on your own. If the scanner shifts to that frequency and you can actually hear calls from the service you selected, confidence in learning to use the unit is high. If you are unsuccessful in entering frequencies, ask to see the unit's instruction manual. Is it easy to understand or more complex than the radio? Look at the unit's literature to see if the company offers an 800 number for technical support for new owners. Exotic scanners may look impressive, but they are of little use during emergencies if you must spend an excessive amount of time programming them beforehand. Ease of operation has its benefits.

A real life example of the previous situation occurred recently while I was visiting a radio store. A customer asked to see some scanners capable of receiving local utilities, and the salesman put three models on the counter to show him. The customer asked for a demonstration of one, but the salesperson could not enter frequencies directly on the unit's keypad. Indeed, both salesperson and customer spent over five minutes trying to get the unit out of memory/scan mode. In final desperation, the salesperson tried a search function scanning from 150 to 800 MHz. The scanner stopped on various activities within a 10-minute period before it ever reached the police band range. A call was eventually heard (police?), but when a button was pressed to stop scanning, the unit returned to a preprogrammed 800 MHz frequency in memory. The perplexed customer suggested that a handheld scanner with fewer buttons might be easier to understand and use. The pair then focused on a neat-looking handheld but alas it, too, required pre-programming a range rather than entering frequencies directly. Soon thereafter, the customer left without purchasing a scanner. Possibly he visited another store or returned to the same store the next day and studied the manuals until he found a unit which was easy to understand and operate.

Once you have access to a scanner's literature and/or manual, you can also check its technical specifications. Here you will find a full list of its frequency coverage and blocked out (not receivable) frequency ranges. You will also find three important parameters: sensitivity, selectivity, and intermod rejection. Sensitivity describes how well the radio can receive weak signals. Most scanners have a sensitivity of 0.5 microvolt. The lower the microvolt number, the better the sensitivity. That is, 1.0 microvolt is a less sensitive receiver and a 0.25-volt sensitivity is a very good receiver. Selectivity describes how well the scanner can reject adjacent frequency/channel interference. The average selectivity of a scanner is 30 kHz. A larger selectivity rating means it may be susceptible to interference from services on nearby frequencies.

Intermod immunity or intermod rejection describes how well the scanner will work in an RF-busy environment. That is, how well it can receive signals from police or other service repeaters across town when used in an area near paging repeaters, fire department transmitters, and other two-way radio services. High intermod immunity is especially desirable in larger cities. Poor intermod immunity means the scanner will stop on false signals such as various bleeps, howls and "doctor call your office" messages. While studying specs and relating them to the scanner, also note if the unit has an external antenna socket and an earphone or audio output socket for connection to a tape recorder. Recording intercepted messages borders on being illegal, but are acceptable provided you are the only one to hear them. An external antenna ensures that you hear fringe area signals without drop-out or being "squelched off" in addition to allowing the scanner to use its maximum range. Here, too, we find the importance of intermod immunity influencing the scanner. If an outside antenna is used for greater range, the receiver is more susceptible to intermod interference. The logical solution is usually determined by trial and error. If you live within a metal structure, an outdoor antenna or small antenna taped to a window may provide sufficient range while avoiding intermod. Alternately, you may want an outdoor antenna for receiving fringe area signals and a smaller indoor antenna to minimize intermod during busy daytime hours. Numerous antennas and preassembled cables or even custom-made cables with connectors to fit your radio are available from dealers nationwide. Additional information on antennas is presented later in this chapter.

As mentioned earlier, scan speed is another high-priority consideration for monitoring numerous services at the same time with a single scanner. High speed scan is also desirable for following trunking operations within the 800-MHz band. As this book is being written, some of the latest scanners will zip from 30 MHz to 1 GHz in less than two seconds. These scanners are also capable of checking 100 to 200 programmed memories within one or two seconds. The faster the scan speed, the greater your chances of hearing brief transmissions of interest. If your scanner is not capable of high speed scan, narrow down your desired monitoring range and program those limits into memory for more rapid "find the action" results. (For example: 156-157 MHz rather than 150-180 MHz.) Read monthly scanning magazines and stay current on the latest scanning accessories. Optoelectronics, for example, produces a receiver that sweeps from 30 MHz to 2 GHz in less than one second and reads out the precise frequencies it finds on a built-in display. The unit includes a speaker output so you can hear the signal as it is received. The unit is different from a conventional scanner in the respect it must be within close proximity (one city block or less) of the transmitter and "locks on to" the strongest signal it receives. When combined with a scanner and used on the road or at a disaster site, it makes an ideal means of spotting on-the-air activities.

Power Sources and Scanners

Certainly the most important consideration to survivalists is how their scanner is powered. A unit that operates solely from 120-volt AC is of little use when conventional power is out. Inverters that change 13.8-volts DC from an auto storage battery to 110/120-volt AC are popular alternatives for emergency use, but avoid small low-cost inverters incapable of handling sufficient current. Look at the back of your scanner or study its manual and note its current requirements. If, for example, it requires .5 amperes while using 120 volts, multiply .5 times 120 to establish 60 watts required. Then, select an inverter capable of handling more than 60 watts for comfortable operation. Inverters are a temporary solution, and should not be considered as the ultimate answer for long-term use. Why? Simply because they convert one energy level to another energy level losing efficiency in the process. They also require inconvenient recharging of the auto's battery.

Through intelligent study and shopping, you can find a scanner capable of direct operation from both AC or DC sources. Ideally, separate power cords will be supplied for home and battery use. One popular trend involves using a home "wall adapter" power supply for the scanner. These units usually have a coaxial socket on the rear panel marked "12-14 VDC input." When connected to a gel cell or similar rechargeable battery, they are usually good for seven or eight hours of monitoring. The shortcomings are remembering to maintain the batteries at full charge for impromptu use and unplugging the wall adapter when the scanner is not in operation. The scanner's on/off switch controls only the unit; the wall adapter draws current all the time it is plugged in.

Figure 5-8 *A novel and lightweight rechargeable power supply for any scanner. This unit will supply 12 volts DC and will power the average scanner for days. Can be recharged from an auto cigarette lighter plug, from a small solar panel, or from a 110 VAC line.*

When permitted by state laws, a mobile scanner has its benefits. Not only can it be used in the car, a rechargeable battery pack (or two) permits use indoors or in the field. Further, small 110/120-volt AC to 13-volt DC power supplies are readily available from radio dealers to power mobile scanners indoors. Serious scanning enthusiasts might consider making a carryall containing two 6 ampere-hour rechargeable batteries, a home power supply and recharger, and a mobile scanner. With clever planning, the unit could be used anywhere on a moment's notice—especially if several types of scanning antennas used with various monitoring situations were included in a side pouch. Use your ingenuity and think survival and emergency situations.

Quite possibly, a handheld scanner gives one the best of all worlds. Newer units are exceptionally compact (they fit in a coat pocket), and include most of the advanced features found in deluxe base scanners. Likewise, some of the basic units are easy to use and some of the more elaborate units are quite complex to program. Again, I suggest trying to enter frequencies or scan ranges before purchasing to determine the ease of use and how well you understand the instruction manual. Important features to look for in a hand scanner are fast-speed scanning, full coverage from 30 MHz to at least 1.3 GHz, an antenna connector, and type of batteries/power used. Most popular scanners use a BNC connector and are sold with a "rubber ducky" antenna. The flexible antenna is good for receiving nearby signals, and accessory antennas can be substituted for monitoring fringe areas. Additionally, magnetic mount or clip-on window antennas can be plugged into a handheld scanner's BNC socket for mobile use.

Some scanners utilize a few "AA" (penlight) 1.5-volt DC batteries for power. A few are supplied with rechargeable batteries. Some of the rechargeable batteries are conventional

"AA" type, others are specifically designed packs you must get from the manufacturer or third party producer. Handhelds that use conventional "AA" batteries are particularly attractive, in that the small energy cells are available from drug stores and variety stores nationwide. Conventional dry cell/carbon zinc batteries have the shortest life span. Alkaline batteries provide a longer life span and more hours of operation. Rechargeable nickel cadmium batteries are quite appealing, because you can monitor for most of a day and recharge them for five to eight hours at night. During emergencies, however, recharging may not be possible. Having a good stock of alkaline batteries on hand is always an advantage.

Many handheld scanners also have an external power jack that allows you to operate them from an external battery or an AC power supply for home use. In this case, the compact unit serves triple functions. Study a considered unit's manual in particular to find if it is possible to power the handheld from an external AC supply while recharging the internal batteries via a wall adapter. In this case, the radio can be used almost nonstop, as needed. Base, mobile, handheld—which scanner is best for you? If in doubt, get one of each! Yes indeed, that arrangement provides monitoring anytime, on-the-spot, at its best! Even the best scanner with the best power supply is of little use if it is not with you when needed. Armed with a good frequency guide and portable scanner, you are equipped to face almost any situation in which you need immediate information. Speaking of frequency guides, the "Betty Bearcat" books sold by dealers nationwide are outstanding. Two top magazines for scanning are "Monitoring Times" and "Popular Communications." Both magazines are published monthly and feature columns on all aspects of scanning and monitoring. They are available on newsstands nationwide.

Scanner Features and Functions

Today's VHF/UHF scanning receivers are a marvel of modern technology, and new models with even more advanced features are being introduced every year. Newcomers may not be familiar with basic scanner terminology, so let's begin with a brief discussion of functions related to a scanner's front panel controls. We will then examine a few presently popular scanners to familiarize you with what's hot on today's market.

The front panel of a typical scanner sports a channel/frequency readout, keypad with controls for volume and squelch, and several pushbuttons for various functions. Exact labeling of controls may vary from model to model, but the basic concepts of each are the same.

Manual Key Places the scanner in manual mode. Each press of the key advances the receiver one channel.

Arrow Keys Arrows change scan or search direction upward or downward in frequency.

Limit Key Sets upper and lower limits in a scan/search range.

Number Keys When in direct frequency mode, enters frequencies. When in memory mode, selects memories or banks of memories.

Scan Key Places the unit in automatic scanning of entered frequencies or memories.

Delay Key Allows the receiver to pause on an active channel/frequency for monitoring.

Lockout Key Locks out individual memories so they will be skipped while scanning.

Priority Key When programmed with a particular activity of interest, pressing this button immediately shifts scanner reception to that channel/frequency. In some scanners, priority also toggles between assigned frequency and tuned frequency.

Direct Key Starts a frequency search up or down from frequency/channel showing in display. Note: similar function as limit key. Different brand scanners use different designations on keys.

Enter Key Enters displayed frequency into a memory.

Clear Key Erases an incorrect entry if pressed before pressing Enter key.

Monitor Key Disables search/scan mode and permits monitoring a specific channel.

Auto Key Permits automatically storing active frequencies found while scanning into unused memory channels.

Weather Key One-touch access to preprogrammed NOAA weather channels.

Mode Key Switches receiving mode between AM, FM, and wideband FM.

Step Key Changes frequency steps in search/scan mode; usually between 5, 12.5 and 50 kHz.

ATT In Key Inserts 10 or 20 db of attenuation in received signal path to reduce interference and intermod.

On/Off Volume Control Turns the scanner on/off and adjusts volume level.

Squelch Control Adjusts the scanner's squelch to eliminate background hiss. Excessive squelch will prevent the unit from receiving weaker signals. Note: The receiver must be squelched in order to scan.

Several controls and connectors are also located on the back of a scanner. Familiarity with each is beneficial.

Antenna Socket Used for connecting a telescoping antenna or outdoor antenna. Typically larger connector is SO-239 and smaller "push and turn" type socket is BNC.

13.8 DC Permits the scanner to be used when powered from an external 12-volt DC source.

External Speaker Used for connecting an external speaker or earphones to a scanner.

Tape/Monitor Output for connecting the scanner to a tape recorder.

Line Out Used for connecting to a line input socket on a tape recorder or PA system. Note: Line level is one volt, whereas the tape recorder level is .001 volt. Connecting a line out to a tape recorder's "microphone in" injects an exceptionally high level into the recorder and produces distortion. Likewise, connecting the" tape out" of a scanner to the "line in" of a recorder produces no recording because the level is approximately 1,000 times too weak.

Reset Pushbutton Resets scanner to default mode and search steps. May clear/erase all frequencies entered into memory. Check your manual before pushing this button!

Understanding and using controls on your scanner is actually easy. Usually all that is required is a little time and patience studying the unit while reading the manual. Remember that no one knows your scanner better than the people who made it, and its manual is your personal assistant/guide to proper operation. Through experience, I have noticed people working with any type of electronic equipment do not always take time to read and comprehend the unit's instruction manual. I take the same unit and the same manual, and "get it going" with no problem. The secret? Look for what you are doing wrong rather than trying to confirm what you are doing right. Read that sentence again. It is your key to success. Try to think like the person who wrote the manual. Follow the steps and duplicate the procedures; then apply the proven technique to the function you wish to use. Once you get the idea, and use the scanner for a few hours, operation will become second nature. Honest! Now let's examine a few of today's popular scanners.

The Scanner Market

AOR Model AR-5000 "CYBER-SCANNER" This is one of the latest and most elaborate base scanners on today's market. It covers from 10 kHz to 2.6 GHz. Its modes of reception include AM, FM (wide and narrow), lower and upper Single Side Band, and CW.

Figure 5-9 *AOR's AR-5000 is a remarkable example of new style "do it all" scanners. Unit covers 10 kHz to 2600 MHz on all modes, all frequencies, and uses micro miniature circuitry to pack a room full of electronics into a small cabinet.*

Figure 5-10 *Unfamiliar with scanning techniques and/ or active frequencies in your area? This Bearcat BC-890XLT finds all the hot spots for you and loads them into memories for convenient review. Unit covers 29 to 956 MHz in 23 bands; some frequency ranges are skipped.*

That is right: This single cabinet unit tunes shortwave broadcast and communications plus all VHF/UHF frequencies. The receiver's design is superb with triple conversion for excellent intermod immunity. The AR-5000 features 1,000 memories divided into 10 banks of 100 channels each. Keeping track of 1,000 memories may seem like an insurmountable task; however, most people delegate banks of memories according to preference. For example: 10 banks may be used for shortwave broadcast station reception, 10 banks for law enforcement agencies, 10 banks for fire departments, 10 banks for weather and ham repeaters, etc. Additionally, the first memory in each bank can be specified as a priority location for instant access. The AR-5000 is supplied with an AC adapter, and also may be powered directly from a 12-volt DC source capable of providing one ampere for each hour's use. The unit's size is approximately 3.5 by 8.5 by 10 inches (H,W,D).

Uniden/Bearcat Model BC-890XLT Scanner This impressive base/portable unit covers 29 MHz to 956 MHz in 13 bands, except the cell phone band is locked out. It has 200 channels separated into 10 banks for convenience in storing services according to use. The BC-890XLT also has an automatic store feature that searches for action and stores frequencies in memory channels. It will scan 16 or 100 channels per second. The unit also includes a weather alert function which detects tones from the NOAA weather service and activates the receiver when emergencies occur. An auxiliary tape output socket and line output socket are included. The BC-890XLT is supplied with an AC adapter or operates from an external 13 VDC source. The front panel's tuning knob is especially appealing for dialing up frequencies when punching in exact frequencies is inconvenient. A tuning knob usually makes a scanner easier to use by beginners.

Uniden/Bearcat BC-220XLT Handheld Scanner The Uniden/Bearcat BC-220XLT is the next best thing to a base scanner. This unit covers several ranges in a "skip range" manner.

Figure 5-11 *This Bearcat BC-220XLT covers the same frequency ranges as the BC-890XLT, but it slips into a coat pocket for go anywhere monitoring of police, fire, weather, hams, etc.*

That is, it receives 28 to 54 MHz, 108 to 174 MHz, 216 to 512 MHz and 806 to 956 MHz. It has 200 memories separated into 10 banks and 10 priority memories. It also features a turbo scan which will skip through 100 memories per second. Additional features include programmable search mode, single punch weather key, and a convenient preprogrammed

Figure 5-12 *Interested in monitoring covert and/or unusual band activities with a handheld scanner? Select a unit like this Bearcat BC-3000XLT. It has continuous coverage from 25 MHz to 1300 MHz FM plus turbo speed scan to catch elusive action.*

service search which permits toggling through police, fire, aircraft and marine bands with ease. The BC-220XLT is supplied with a rechargeable battery pack, AC adapter/battery charger, and an optional cable for powering/charging the unit from an auto's cigarette lighter socket is available from dealers.

Uniden/Bearcat BC-3000XLT Handheld Scanner The BC-3000XLT is an elaborate handheld scanner covering 25 MHz to 1.3 GHz, with the exception of cellular telephones. The unit sports 400 memories divided into 20 banks plus turbo scan which zips through 100 channels/memories per second. Particularly appealing is the unit's automatic store function which searches frequencies and stores active ones into available memories. It will also automatically sort memories for optimized scanning. In other words, you can simply set the unit to scanning and return a few minutes later and review any activity found. The BC-3000XLT is supplied with a rechargeable battery pack and AC adapter and charger. An optional automobile power cable is available from dealers.

Uniden/Bearcat SC-150 "Sport Cat" Scanner An impressive handheld scanner covering 28 to 54, 108 to 174, 406 to 512, and 806 to 956 MHz. This unit features 100 memories and 10 priority channels, in addition to preprogrammed band search for finding action in police, fire, aircraft and marine bands. The SC-150 also features one-touch weather reception, turbo scan, and one-touch access to memories. This particular unit is a good example of nice looks overwhelming ease of operation. It looks simple, but it is actually complex to learn. Truly, one "cannot judge a book by its cover!" The SC-150 is supplied with rechargeable battery, and an AC adapter/charger.

Figure 5-13 *Classy looking Bearcat SC-150 is available in black or yellow case and covers popular bands between 29 and 512 MHz. SC-150 offers good performance and economical cost, but learning how to program it for specific frequencies can be challenging.*

Optoelectronics XPLORER Although not a scanner per se, this unusual unit deserves mention. The XPLORER is a high speed, "near-field" receiver that sweeps the range of 30 MHz to 2 GHz in less than one second. It locks onto the strongest signal received, displays its frequency and strength, and outputs its audio on a built-in speaker. By pressing a button, you can lock out strong frequencies and continue searching for weaker frequencies. You can also program your scanner with frequencies found by the XPLORER. The unit also includes the decoding of DTMF (touchtone) and CTCSS (subaudible) tones used in codes. Additionally, as the XPLORER finds active frequencies, it can store up to 500 of them in its memories. The term "near-field receiver" means the XPLORER must be close to transmitting stations to receive them, typically within a one-half block range. When combined with a handheld scanner, the combination may be an ultimate monitoring setup for tuning in all action at a crisis site.

Additional information on all previously discussed scanning equipment (AOR, Uniden/ Bearcat, XPLORER/Optoelectronics is available from shortwave radio dealers nationwide.

Numerous other models of base, portable/ mobile, and handheld scanners are available on today's market. Our brief sampling shown here gives you a general idea of what they look like and what they do.

Figure 5-14 *Want to know what frequencies are being used by that SWAT truck in front of you, that handheld on the corner or a tower you are passing? Just switch on an Optoelectronics "Xplorer". It locks onto frequencies, displays them, stores them in memories, and monitors related transmissions on its speaker. Unit must be in close proximity to signals to find and monitor them.*

External Antennas for Scanners

Looking for a good way to increase the reception range or distance your scanner will cover? Need to hear those weak communications on the scene or across town? The answer is to add an external antenna to your base, mobile or handheld scanner. Indeed, any radio, transceiver, or scanner's ability to "hear better" is directly related to its antenna. Further, longer does not always mean better. The antenna should be resonant (cut to a precise length) for a desired range of frequencies. The pull-up antenna supplied with most scanners is little more than a semi-resonant rod, and "rubber ducky" type antennas are simply a wire spring enclosed in a plastic shield. Their performance is usually acceptable, but not optimum. Substituting a commercially-made or a homemade antenna can open a new world of scanning for you. Do not just take our word for that fact; try it for yourself and see. If you are unfamiliar with setting up outdoor antennas, preassembled cable sets are available in various lengths for easy "plug and play" convenience. Simply plug one connector into the antenna, thoroughly weatherproof it with sealant, and connect the other end to your scanner. Tip: When leaving your scanner during storms, unplug the antenna to avoid "front-end" damage from lightning in the vicinity. Three types of connectors are commonly used with scanners. An SO-239 socket and a PL-259 plug (which is large and screws on like the ones on CB radios). The BNC connector (which is slightly smaller; pushes on and turns to lock). These types of connectors are used on laboratory equipment and scanners. Finally, "F" connectors similar to the types used with cable TV hookups may be used. These small connectors screw on like a cable TV antenna. Now let's examine two commercial antennas.

Diamond D-130J This discone antenna, shown in Figure 5-15 features an exceptionally wide bandwidth and full 360-degree coverage. Various elements on the antenna are tuned to frequencies between 25 and 1300 MHz, allowing it to receive signals on all frequencies within that range. The antenna stands approximately four feet tall, and has an SO-239 jack to accept a PL-259 plug. The cable for connecting the antenna to a radio is not supplied and should be purchased separately. The discone design is well known and respected for good performance over an exceptionally wide frequency range. As a result, it makes a very good scanning antenna.

Grove Scan Beam This antenna is designed to pick up more distant stations, and you must point it in the precise direction from which you wish to receive signals. Many people install beam antennas of this nature and rotate them with a small TV antenna rotor. The Grove Scan Beam utilizes multiple resonant elements to cover 30 to 50 MHz, 108 to 136 MHz, 136 to 174 MHz, 225 to 400 MHz, 406 to 512 MHz, and 806 to 960 MHz. The antenna is fitted with an "F" connector. The connecting coax cable must be purchased separately. The antenna is 8 feet tall by 5 feet long.

When installing any type of outdoor antenna, always stay clear of power lines. Quite surprisingly, more people have been killed by electrocution while installing antennas on windy days and trying to catch them as they fall onto power lines than any other type of electrical accident associated with radios. Never install any type of antenna near any power lines. Ideally, an outdoor antenna should be mounted as high as possible. There are logical exceptions in that statement. First, avoid mounting an outdoor antenna so it is the highest object around. Let

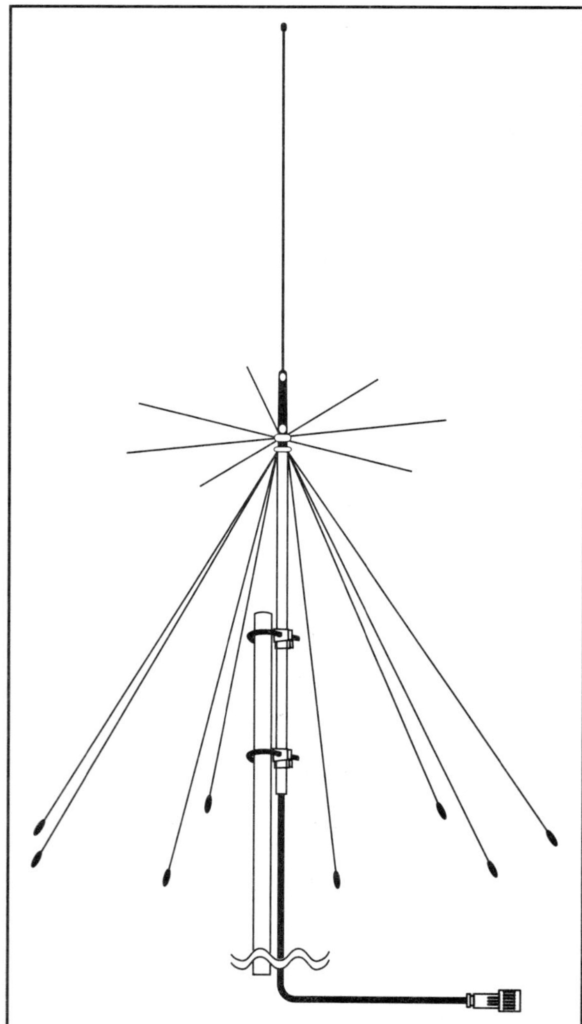

Figure 5-15 The Diamond D-130J Discone Antenna. Item is 4' tall and covers 25 to 1300 MHz range well. Discones are superb for omnidirectional reception.

some other structure, tree, or power pole be the scapegoat for lightning. Additionally, tall antennas attract curious eyes and they may be prohibited in congested residential areas. If your dwelling has antenna restrictions, consider mounting the antenna in the attic.

Like to try your hand at making your own scanner antenna? Check out our simple design shown in Figure 5-17. Assuming you have a 15- to 25-foot length of 50-ohm/RG-58 cable with a connector on one end to fit your scanner, measure 18 inches from its "far end," then carefully slit the outer jacket and remove it to reveal the inner braid/shield. Push the shield back slightly from the far end to provide some slack and make an opening in the braid. Next, carefully feed the insulated center conductor out through the opening to produce pigtail leads. Then fold the braid back over the coax and tape it in place so that the insulated center conductor remains. Tie a string or other nonconductive support to the top of the antenna (center conductor), then hang it out of a window. Route the cable to your scanner, plug it in, and enjoy scanning supreme.

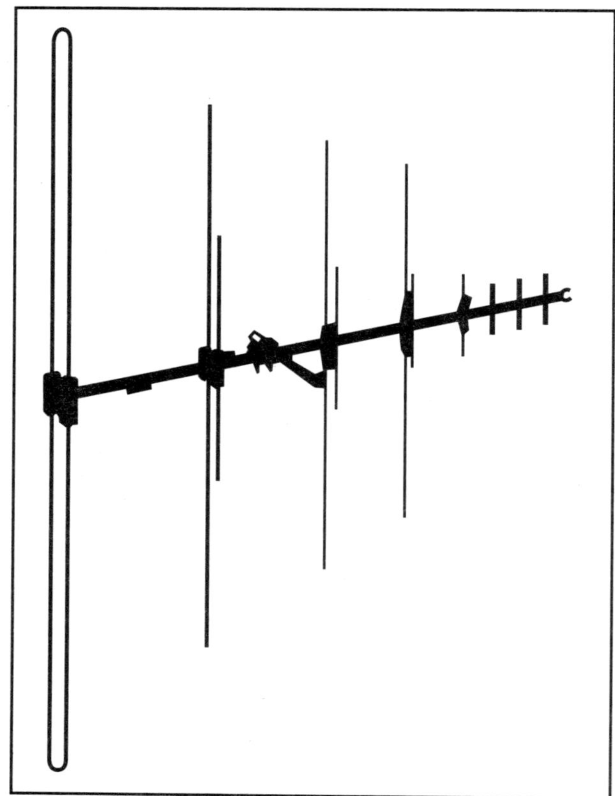

Figure 5-16 The Grove Scan Beam is especially useful for receiving distant stations or weak across town handheld transceiver signals. Discussion in text.

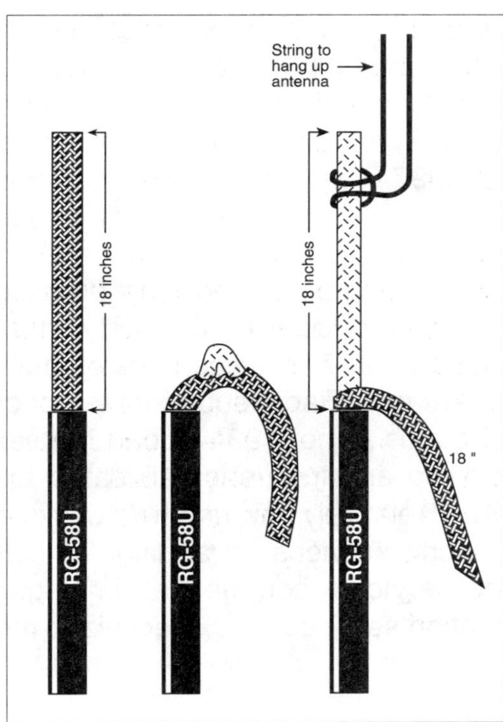

Figure 5-17 *An easy to make antenna for scanner reception in the 100-170 MHz range. If desired, whip and shield sections can be trimmed to 9" for reception in 300-500 MHz range.*

Many people report poor reception when using scanners in automobiles. Naturally! The little rubber ducky antenna is enclosed within a metal shield. What to do? Valor has a neat little accessory you simply slip it onto an auto's window, mount your "rubber ducky" antenna on the clip so it is outside the car's window, then route the cable to your scanner. This simple solution certainly beats holding the scanner out a window! Additional antenna ideas are absolutely endless. Indeed, experimenting with antenna ideas and designs is a world all its own—and far too extensive for this introductory manual.

Our modern age is filled with numerous uncertainties and need-to-know situations erupting with little or no forewarning. Scanning local area/high-band activities provides vital "what is happening" insight without sacrificing personal security during crucial times. When away from home base, a portable scanner can provide authoritative information on an immediate basis. Investing in a good scanner is a good survival tactic.

Figure 5-18 *A mobile magnetic antenna mount with coax cable. Unit shown with several antenna lengths (sections) to cover VHF and UHF frequencies.*

ATTENTION **ATTENTION** **ATTENTION**

ARLB046 ARRL studies new eavesdropping bill

> From ARRL Headquarters Newington CT August 8, 1997
To all radio amateurs

The ARRL is closely studying another bill introduced in Congress to beef up prohibitions against electronic eavesdropping. The bill, HR 2369, was introduced by Rep Billy Tauzin of Louisiana. Dubbed the Wireless Privacy Enhancement Act of 1997, it has scanner enthusiasts and equipment makers worried and could affect some Amateur Radio equipment. If passed, it would—among other things—amend the Communications Act of 1934 to ban the sale of scanning receivers capable of receiving transmissions on any frequency allocated to any Commercial Mobile Radio Service (CMRS). The CMRS is a relatively new umbrella designation of subscriber-based radio services that act like telephone services. In addition to cellular telephone, such services include commercial paging services, commercial air-to-ground services, offshore radiotelephone, personal communication services, and specialized mobile radio services.

HR 2369 would prohibit receiving, divulging, publishing or using any intercepted transmission, and subject violators to substantial fines or prison terms. It also would make it illegal to modify equipment so that it may be used to unlawfully intercept or divulge radio communications. The FCC would be charged with investigating complaints and enforcing the stiffer regulations.

As currently drafted, the bill appears to affect equipment available to scanner enthusiasts, hams who use scanning transceivers to receive out-of-band, and hams who use out-of-band capability for volunteer work. It would not affect ham frequencies, per se. The League's Legislative and Public Relations Manager, Steve Mansfield, N1MZA, says the ARRL is studying the bill to determine its long term implications for ham radio and ham gear. The ARRL has contacted Tauzin's office to express its concerns, and Mansfield says the League "will continue to work with members of Congress to have the bill modified to reflect the needs of the Amateur Radio community."

The Tauzin bill comes fast on the heels of very similar, but less-stringent, legislation proposed by Rep Edward Markey of Massachusetts (see The ARRL Letter, Vol. 16, No 29). The ARRL has met with Markey's staff to discuss the negative implications of HR 1964 for Amateur Radio. That bill was not given much chance of passage.

An incident last year in which House Speaker Newt Gingrich's cellular telephone conversation was illegally intercepted, taped and published by the media prompted calls in Congress for stronger anti-eavesdropping legislation.

Note: Do not wait too long to obtain your scanning receivers and equipment—it looks like they might be highly regulated in the near future!

CHAPTER SIX

Citizens Band Radio

Logical survival-oriented thinking dictates being prepared for any type of abnormal situation and for maintaining familiarity with various means of communication to fit associated needs. Sometimes those needs involve monitoring local or international information sources. Sometimes it involves using disciplined or personal communications with a specific group. Other times, survival preparedness encompasses communicating with the public at large. Considering those ground-floor facts, the most popular and readily available form of two-way communication on a large scale is Citizens Band radio.

The Federal Communications Commission established our present 11-meter/27-MHz Citizens Band in 1958. Its purpose was designated as short-range communications (155 miles maximum, 10 to 20 miles typically) through a number of specified channels to serve small businesses and families, providing a way to stay in touch between homes/bases and autos. Initially, users were required to obtain a license and call letters issued by the F.C.C. to operate a CB transceiver. Issuing licenses became an insurmountable task, so formal licensing was abandoned soon thereafter. Today, you can purchase CB radios, antennas, and accessories from a variety of stores nationwide and go on the air immediately without a license or call letters. The Citizens Band is, however, still governed by the F.C.C. Rules, part 97, subpart D. These rules cover types of communications permitted and restricted, power and antenna limitations, and equipment stipulations. They designate the Citizens Band as a two-way service for personal or business activities, usable throughout the 50 states, plus Puerto Rico and the Virgin Islands.

After opening on 26/27 MHz in late 1958, CB grew into an absolute craze during the mid-1970s. The CB fad continued several years, and illegal activities like using high power, "shooting skip," using "sliders" to operate between channels or out of the band became prominent. The CB boom began to calm during the early 1980s, and it evolved into a more useful two-way service by 1990. It also emerged with a total of 40 channels, more technically advanced equipment, and a most commendable nationwide emergency monitoring network (REACT) on Channel 9. Today, CB is experiencing a vigorous comeback. It is used for both public and private communications of all types and often serves a viable purpose during on-the-road emergencies. Terms such as "good buddy," "cotton picker," and "Tijuana Taxi" are outdated. Continuing is CB's use of "10 codes" and unique jargon, and occasionally harsh

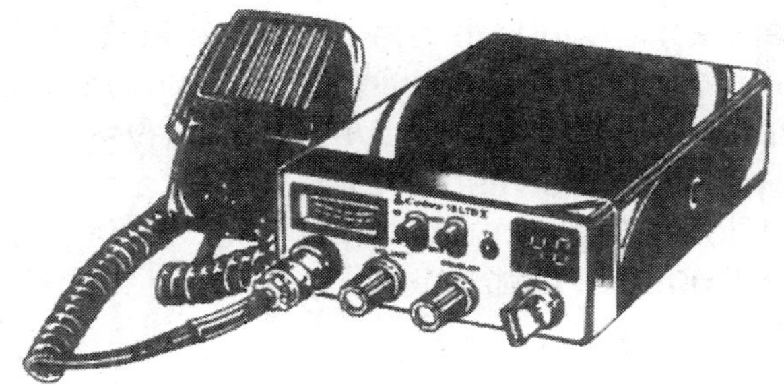

Figure 6-1 *CB radios are easy to use and quite handy for communicating with other members of society during all types of normal or unusual situations or emergencies. Units are available in mobile, base station, and handheld versions.*

Figure 6-2 CB is one of the most popular and often-used communications media on open and/or congested freeways and byways. It is akin to a continuous information exchange among travelers and commuters everywhere.

language. Over 50 million CB radios are in use throughout the US; it is akin to a gigantic party line or intercom among truckers, four-wheel travelers, RVers, boaters, and people of all backgrounds. Indeed, CB is today's most common means of staying attuned to what is happening on freeways, highways, and byways. CB is also inexpensive; a typical mobile unit for an auto may cost as little as $50 to $60 and there are no monthly usage charges.

CB: An Asset or an Aggravation?

The general public is a vast cross section of personalities and interests. Put a microphone in the hands of one out of five, and you have a good overview of the CB world concept. Dozens of users share a limited number of channels, so a certain amount of interference and confusion is natural. Although interference is usually unintentional, it is the main factor limiting the communications range on CB. If a channel is clear, reaching out 10 to 20 miles or farther is normal. If a channel is congested, range may be confined to 4 or 5 miles. Coarse language, usually from inconsiderate users, is another aspect to be considered or ignored. Assuming the previously mentioned idiosyncracies are not overly offensive to you personally, CB offers several benefits, especially when traveling.

With a majority of the population using, owning, or having quick access to a CB radio, its appeal for contacting someone from any location is an established fact. The "someone" first contacted may be a role model citizen or an undesirable individual; let your survival instincts and sense of good judgement be your guide. For example, you would not want to openly announce on a traveler's channel (19) that you were alone, carrying cash, and stranded on a back road without a backup plan if scoundrels responded to the call.

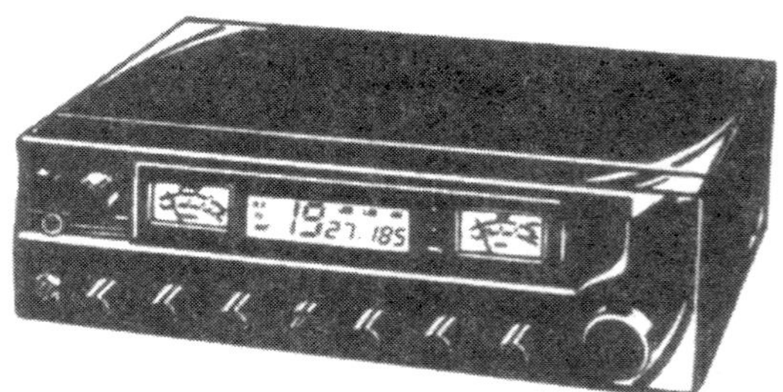

Figure 6-3 Basic style CB transceivers are quite low in cost and high in emergency preparedness value. Truly, it is better to have one and not need it than to need one and not have it.

Figure 6-4 What is happening beyond your line of sight? Tune in nearby truckers on channel 19 for "balcony seat reports."

If you were one of thousands evacuating a coastal area prior to a hurricane making landfall, CB would be a good way to communicate with other evacuees. A CB radio would also prove useful for getting into bumper-to-bumper traffic moving out of the area. If a "pit stop" along the way was unavoidable, communicating to other drivers your intention to make a courteous lane shift to exit or return (and perhaps a report from the last person stopping in that same rest area) could also be handled via CB.

Have you ever been marooned in a two-hour traffic backup on a freeway? Have you noticed that few professional truckers are found in such two-lane parking lots? Why? Chances are good that they compared notes with other truckers coming from the opposite direction, via CB channel 19, and took an alternate route. Truckers sit high above auto traffic, can see quite a distance ahead and behind, and will usually give you accurate information when asked in a sincere manner. Would you not prefer using a CB radio for information to avoid a two-hour delay rather than using a cell phone to report your arrival at a destination would be two hours late? Another point in each of the prior hypothetical cases a high-power CB setup capable of long range was not necessary. All related communications were within a one or two mile area, and could have been conducted with a portable handheld CB.

Now think a bit further. Assuming you are a licensed radio amateur and also have a VHF FM mobile transceiver in the auto, you can check back through repeaters behind you or call through repeaters ahead for lodging information. You could also maintain communications with emergency nets in cities along the route to acquire dependable and accurate "what's happening" details as you travel. Add an HF amateur radio transceiver in the auto, and it becomes a fully mobile communications center capable of worldwide range. CB and amateur radio both have their unique applications; each serves a specific purpose and each has its benefits.

More applications for CB also warrant recognition. If you awoke to discover a surprise snow or ice storm was closing main roads, listening on CB to ascertain conditions around your area (and make emergency phone calls for others) would be wise thinking. Neighbors would probably be using CBs also, providing more insight into nearby conditions. In fact, CB is used quite often in neighborhood watches and as a community hot line in rural areas. Many police and highway patrol vehicles even have a CB monitoring channel 9, 19, or a locally-adopted "assistance channel." Boaters use CBs on lakes and near shore lines. Even the Coast Guard listens on channel 9 for emergency calls. Why? Once again, CB is the most accessible means of communication for the public at large.

Notes and Tips on Using CB

Although little knowledge of professional communications techniques is required to operate a CB radio, being familiar with a few pertinent details and rules of the (CB) road help ensure successful use. First, let's recognize the two modes of (voice) communication and their areas/channels of use on CB. One, **AM** is utilized on channels 1 through 23 and is associated with the less expensive and the most widely used CB radios. They are restricted to 4 watts output power. Two, **SSB** or Single Side Band is mainly used on channels 24 through 40. Equipment is more expensive and operators are more specialized. Output power is limited to 12 watts P.E.P. Conduct and activity are more akin to Amateur Radio than CB. "Sidebanders" on CB go by quasi-call letters (usually associated with their sideband group or club) rather than "handles" like the ones used on AM. Alternately, a new sidebander might adopt using the letter "K" and their initials followed by the last digits of their zip code until they acquire an official club number.

When tuned in on a regular AM CB, SSB is not understandable. It sounds like a duck quacking. When AM is tuned in on an SSB radio, howls are present until "zero beat" with the radio's clarifier, or until the SSB radio is switched to AM mode. When two or more AM signals occupy the same channel (common in channels 1-23 range), heterodynes or "howls and squeals" result. As activity slows, heterodynes/"howls" disappear. Now let's focus on some specific channels and activities.

Figure 6-5 *REACT® logo signs along highways and in various locales indicate nearby volunteer members are monitoring channel 9 to assist people with emergency communications.*

Channel 9 is officially designated for emergency use and traveler's information. Members of REACT (Radio Emergency Associated Citizens Team) located throughout the United States monitor channel 9 on an almost continuous basis. Highway signs indicating various REACT groups monitoring channel 9 are also seen in many areas. Look closely in residential areas and you may spot more REACT/channel 9 signs. Although there is no assurance you will be able to summon REACT assistance in every locale, the possibility is quite high. Affiliation with REACT, incidentally, is a service which members take quite seriously. Indeed, REACT typically handles over 100,000 calls for help and assistance each year. Want to join or get more details? Contact REACT, P.O. Box 998, Wichita, KS 67201 (telephone 316-263-2100).

Channel 19 is the primary monitoring channel for truckers and travelers nationwide. Continuously busy with traffic information, chitchat and small talk of all kinds. Hotter than a firecracker during abnormal road conditions.

Channel 10 and 11 handles the overflow of communications from channel 19. It stays busy with travelers chitchatting and comparing notes on roads, rest stops, weather, etc.

Channel 13 is the RVers and boaters hangout. Retirees and "full timers" traveling and living in motor homes gather or exchange notes via channel 13 when passing on the road. Channel 13 is also used by boaters on lakes or near shore in coastal areas. Boaters and mariners may also use marine transceivers on maritime channels for specific marine communications while relying on CB for general public communications.

Additional Channels Between 1 and 23 may be designated for community use in rural areas or for a neighborhood party line used in metropolitan areas. Listen to several channels and take cues accordingly. Other 1-23 channels are typically used on a first come, first served basis.

Channels 23/24-40 are primarily used by "sidebanders." Clean operating procedures are observed; and the use of harsh language, slang terms and "handles" is avoided. If you wish to join in, brush up on the "Q" codes used in Amateur Radio before making a poor first impression on SSB.

Conversing on CB involves a fair amount of jargon, but it also includes plenty of plain language. Avoid being a "Smokey and the Bandit" imitator: Half of the movie was hoopla and half withered out a decade ago. Maintain a direct and upbeat attitude. Understand that most CBers are good folks

POPULAR CITIZENS BAND 10-CODES

10-1	Receiving poorly	10-34	Trouble at this station
10-2	Receiving well	10-35	Confidential information
10-3	Stop transmitting	10-36	Correct time is
10-4	ok, message received	10-37	Wrecker needed at
10-5	Relay message	10-38	Ambulance needed at
10-6	Busy, stand by	10-39	Your message delivered
10-7	Out of service, leaving the air	10-41	Please turn to channel
10-8	In service, subject to call	10-42	Traffic accident at
10-9	Repeat message	10-43	Traffic tie-up at
10-10	Transmission completed, standing by	10-44	I have a message for you
10-11	Talking too rapidly	10-45	All units within range please report to
10-12	Visitors present	10-50	Break channel
10-13	Advise weather/road conditions	10-60	What is next message number?
10-16	Make pick up at	10-62	Unable to copy, use phone
10-17	Urgent business	10-63	Net directed to
10-18	Anything for us?	10-64	Net clear
10-19	Nothing for you, return to base	10-65	Awaiting your next message/assignment
10-20	My location is	10-67	All units comply
10-21	Call by telephone	10-70	Fire at
10-22	Report in person to	10-71	Proceed with transmission in sequence
10-23	Stand by	10-77	Negative contact
10-24	Completed last assignment	10-81	Reserve hotel room for
10-25	Can you contact	10-82	Reserve room for
10-26	Disregard last information	10-84	My telephone number is
10-27	I am moving to channel	10-85	My address is
10-28	Identify your station	10-91	Talk closer to mike
10-29	Time is up for contact	10-93	Check my frequency on this channel
10-30	Does not conform to FCC rules	10-94	Please give me a long count
10-32	I will give you a radio check	10-99	Mission completed, all units secure
10-33	EMERGENCY TRAFFIC/MESSAGE	10-200	Police needed at

Figure 6-6 *Popular 10-codes used on CB. Keep a copy by your CB radio for quick reference.*

and do not interfere with communications among others intentionally. Avoid jumping on a busy channel and screaming "Break, break!" into the mike. Instead, find a quieter channel, politely ask if anyone is around or monitoring, and state your (self-assigned) "handle." Think courtesy.

As additional guidance for getting started, a list of the most used 10-codes is included in Figure 6-6. Keep them handy, and make a few "practice contacts" on CB so you will be familiar with the service before it is needed. Good luck on CB, and may your emergencies be few!

Tuning in CB on Shortwave Receivers

The Citizens Band is situated on 26/27 MHz, so it falls within the tuning range of most full-coverage shortwave receivers. Assuming that shortwave radio exhibits good sensitivity (ability to copy weak signals) on these upper frequencies, it can prove quite useful for monitoring activities and gathering information (called "reading the mail" in CB jargon). You will not have the ability to talk back to stations, but that may actually prove advantageous, particularly if you desire to maintain a low profile. If the shortwave receiver has memories, you can even store CB channels active in your area into adjacent memory slots for instant recall. Such an arrangement is quite clever, in that you can listen to several channels almost simultaneously.

Inexpensive and older model shortwave radios often drop off in sensitivity on their "D" band or highest frequency range, so check out performance before declaring an unfamiliar receiver "ready for monitoring." Possibly, substituting a simple homemade antenna rather than using the receiver's built-in or telescoping antenna will improve 26/27-MHz reception. A nine-foot or eighteen-foot wire connected to the receiver's antenna terminal usually works well. Longer does not mean better, however. Wires over eighteen feet long are resonant on lower shortwave frequencies. A channel versus frequency list for tuning in CB is shown in Figure 6-8. Remember channels 1 through 23 (26.965 MHz-27.255 MHz) use AM mode and channels 24 through 40 (27.235 MHz to 27.405 MHz) use SSB primarily and AM occasionally. Good listening!

Figure 6-7 *Monitoring CB channels on a shortwave radio is an easy and convenient way to stay informed in your area. Frequency shown (27.065 MHz) corresponds to channel 9.*

The CB Transceiver Features and Functions

How do you intelligently shop for a CB radio? Some people judge a unit according to its price and knob count: If the cost is low and it has a channel selector knob plus volume control, they are satisfied. Some people are influenced by modes of operation. AM-only units are the least expensive. Combination AM/SSB units are more expensive and more elaborate. Others are interested in a unit's size, cosmetic

CITIZENS BAND CHANNELS AND FREQUENCIES			
CHANNEL	FREQ.	CHANNEL	FREQ.
1	26.965	21	27.215
2	26.975	22	27.225
3	26.985	23	27.255
4	27.005	24	27.235
5	27.015	25	27.245
6	27.025	26	27.265
7	27.035	27	27.275
8	27.055	28	27.285
9	27.065	29	27.295
10	27.075	30	27.305
11	27.085	31	27.315
12	27.105	32	27.325
13	27.115	33	27.335
14	27.125	34	27.345
15	27.135	35	27.355
16	27.155	36	27.365
17	27.165	37	27.375
18	27.175	38	27.385
19	27.185	39	27.395
20	27.205	40	27.405

Figure 6-8 *Channel versus frequency guide for tuning in CB activities on a shortwave receiver. Discussion in text.*

appearance, and ease of operation (user friendliness). As additional guidance to studying and understanding CB transceivers, let's examine some front panel controls and features found on CB radios. All of the following controls may not be included on your particular unit They are described here to ensure you have comprehensive guidance.

Channel Selector and Display Most modern CB units feature a rotary knob and backlit LCD panel. If the knob has smooth action and the display has two or three levels of brightness, it is in the "very good" category.

Squelch This control adjusts (usually clockwise) until background noise/hiss on a quiet channel disappears. Do not increase squelch beyond that point unless you wish to restrict the receiving range. Any station coming in on the channel with a signal stronger than the background noise will "break the squelch" and you can copy it at normal volume. Occasionally readjust the squelch to fit channel activity.

Delta Tune or Clarifier is used to shift the receive frequency ever so slightly and improve reception of a desired station on a crowded channel. Normally, it has a center indentation or slot for "no change in frequency."

RF Gain allows you to adjust the overall sensitivity of the unit's receiver. When reduced, it minimizes "splatter" and interference from stronger stations on nearby channels. When increased, it maximizes the receiving range. It is more common on SSB units than on AM units.

Noise Limiter or Noise Blanker Used to minimize or remove ignition noise ("popping") when mobile. It may also be beneficial in minimizing band static. Sometimes an adjustment control is also included. Set it only high enough to minimize noise. Excessive limiting/blanking can introduce "cracking" and distortion on strong signals. Tip: If ignition noise is excessive, try alternate adjustments of the RF Gain and the Noise Blanker.

Tone Control permits varying received audio for optimum bass and treble response. Occasionally useful, also, in minimizing noise.

S Meter/Power Output Indicator Shows relative strength of incoming signals and power output on transmitted signals. It may also include a highly desirable SWR indication. Signals below S9 are usually several miles away. Signals above S9 are nearby. A signal "pinning the meter" at maximum is probably coming from another CBer within eyesight. SWR relates to the amount of power going to your antenna and the amount being reflected from the antenna back to the CB transceiver. A 1:1 ratio is ideal. A 1.5:1 ratio is typical, and a 2:1 ratio is the highest value you should consider acceptable. SWRs above 2:1 usually indicate problems in the antenna, transmission line (coax cable), or end connectors/connections.

PA/CB Switch By connecting an outdoor-type speaker or paging trumpet to a rear socket, this switch allows you to use the CB radio as a public address system. Use it with discretion. Its use may be an infringement of municipal codes in some cities.

Up/Down Buttons A convenience for quickly selecting channels. It is handy for monitoring.

WX Button Used in newer CBs to instantly access NOAA weather service broadcasts in the 162-MHz range.

Ch 9/Ch 19 Button(s) Provides instant access to emergency and travel channels.

Now let's look at a cross-section of CB radios for additional informational benefit.

The CB Transceiver Market

Cobra CB Radios This brand is one of the most well-known and respected names in high performance CB equipment. Construction and operation are exceptional, and the resale value is good. The Cobra line ranges from economical to elaborate, fitting every need and budget.

One of Cobra's economical AM CB transceivers with high marks in popularity is the model 19DX shown in Figure 6-9. This compact unit covers all 40 channels, has an LED-type signal strength meter, a transmit LED indicator, and single-button access to channel 9. The price is under $60.

Next is Cobra's model 29LTDW shown in Figure 6-10. This unit sports an illuminated S/Power Output/SWR/Modulation meter, Noise Blanker, RF Gain Control, and Transmit and Receive LED indicators. Particularly appealing is the 29LTDW's reception of seven NOAA weather channels in the 162-MHz range. The unit is also equipped with a tone decoder and warning system for responding to NOAA weather alerts.

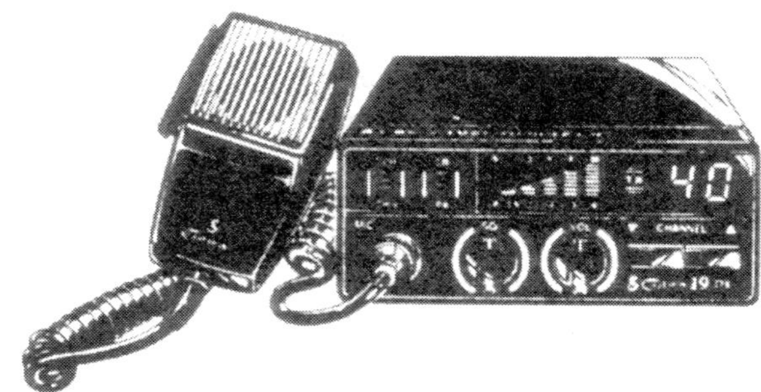

Figure 6-9 *The Cobra Model 19DX reflects outstanding value in a contemporary style CB transceiver. Unit is economically priced and well built.*

Moving to Cobra's top-of-the-line CB, model 2010GTLWX is shown in Figure 6-11. This AM and SSB transceiver features dual meters for monitoring output power, incoming signal strength, SWR, and modulation. The display between the meters indicates both channel numbers and related frequencies in MHz (e.g., five digits, 27.195). The unit also has RF Gain Control, Noise Limiter, clock, and much more. Price for going first class is approximately $400.

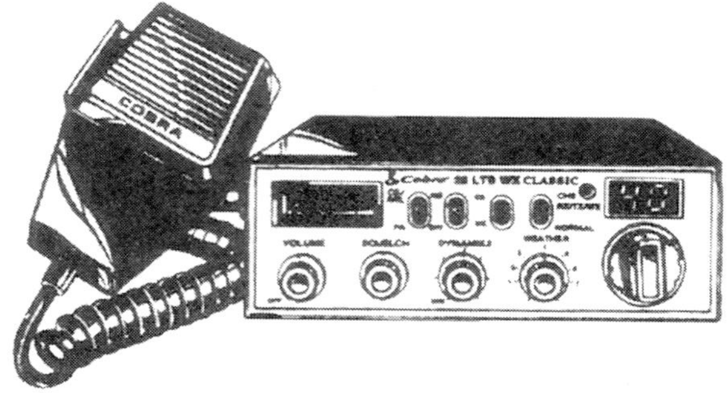

Figure 6-10 *Classic style Cobra 29LTDWX CB unit is loaded with always beneficial features, plus it receives 7 NOAA weather channels. The 29LTDWX makes an excellent traveling companion.*

Maxon CB Radios Another familiar name in CB radios is Maxon. This company produces a quality unit at a reasonable cost. Their model MCB-45W is shown in Figure 6-12. It has separate AF (volume) and RF (sensitivity) controls, LCD channel and "S"/power readout, noise blanker, and a single button call-up of emergency channel 9. The MCB-45W also receives seven NOAA weather channels and three marine channels, making it beneficial for mobile operation and boating. Maxon is sold through electronics stores nationwide.

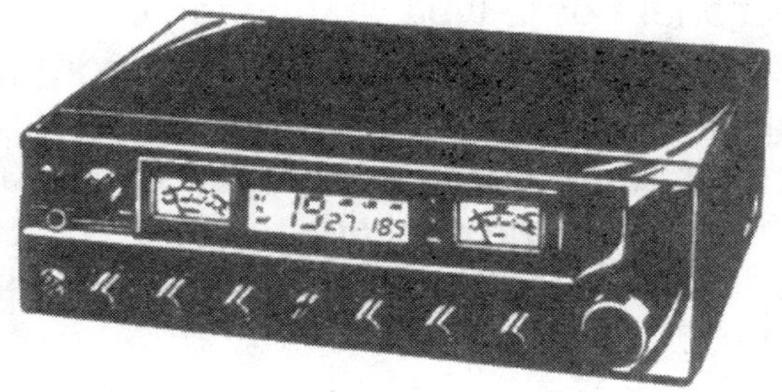

Figure 6-11 *First Class is the only way to describe Cobra's deluxe 2010GTLWX base CB transceiver. Unit operates AM, SSB, receives NOAA weather, and sports a wealth of features.*

Uniden CB Radios This well-known manufacturer of Bearcat scanners also produces a full line of CB radios that are available from suppliers of electronics goods nationwide. Uniden's popular model SSB CB unit is the PC-122XL shown in Figure 6-13. This compact transceiver sports an LED signal indicator, RF Gain control, Squelch, Clarifier and Noise Blanker. It is priced under $150. Uniden manufactures an extensive line of CBs. Check with a local dealer for details.

Figure 6-12 *The Maxon MCB-45W is a deluxe AM CB radio with NOAA weather reception, plus it receives 3 of the most monitored marine channels.*

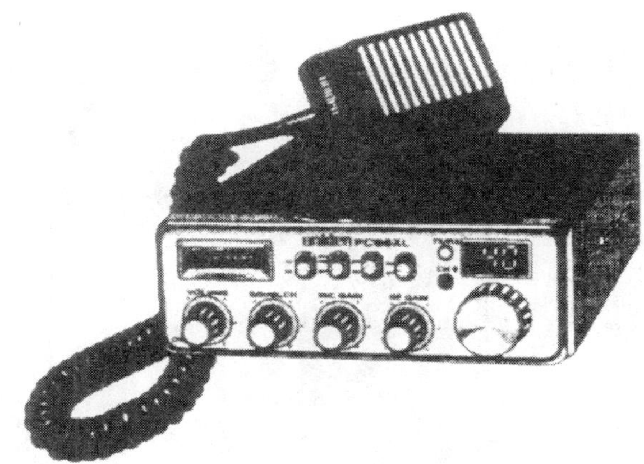

Figure 6-13 *Representing economical cost in an SSB CB unit is Uniden's PC-122XL. Unit has all basic features and functions, yet it is priced comparable to a mid-range AM CB unit.*

Radio Shack CBs Probably the most well-known name in CB and consumer electronics is Radio Shack. Their CB units range from simple to fancy, are competitively priced, and are available in cities large and small nationwide. Surveying Radio Shack's full line is simply impossible.

Handheld CB "Talkies" The popularity of pocket-size CB transceivers is increasing quite rapidly and with good reason. They are ideal for on-the-spot communications. Although small, modern handheld CBs deliver surprisingly good performance. Their range is usually determined/limited by the short "ducky" antennas used with handhelds. Pick one with a top-mounted BNC socket, connect a good mobile or base antenna to the handheld, and it will work as well as a mobile unit.

CB Antennas

As with any type of communications equipment, the antenna connected to it makes the big difference in coverage range. Connect a mediocre antenna to a top-notch CB unit, and the setup's range will likewise be mediocre. Connect a high-gain antenna to a basic or mediocre CB unit, and it will reach out like a champ. How can you visually determine if an antenna is mediocre or marvelous? Generally speaking, vertical antennas up to 8 feet in height deliver average performance while taller versions are noticeably superior in performance and range. Directional beam-type antennas with 2 or 3 elements perform comparable to a high-gain vertical, while beams with more than 3 elements and long crossbeams, pump out a terrific signal in the direction they are pointed. A vertical is much easier to handle and install

Figure 6-14 *Tall vertical antennas like the Valor Pro-Am PS27 are well known for an unobtrusive appearance, omni-directional coverage and strongly radiated signal. This item makes a 5-watt CB unit sound like 40 watts.*

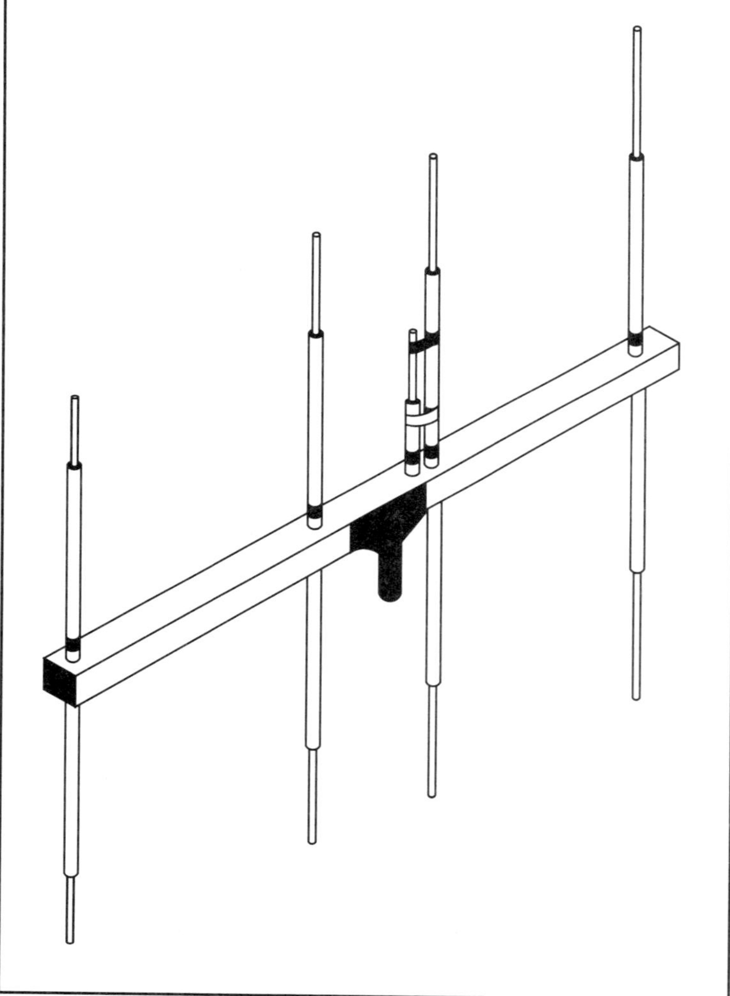

Figure 6-15 *Beam antennas like the popular Jo Gunn JG-4 shown here are ideal for communicating with stations in a selected direction while minimizing interference from other directions. Antenna also boosts effective radiated power significantly.*

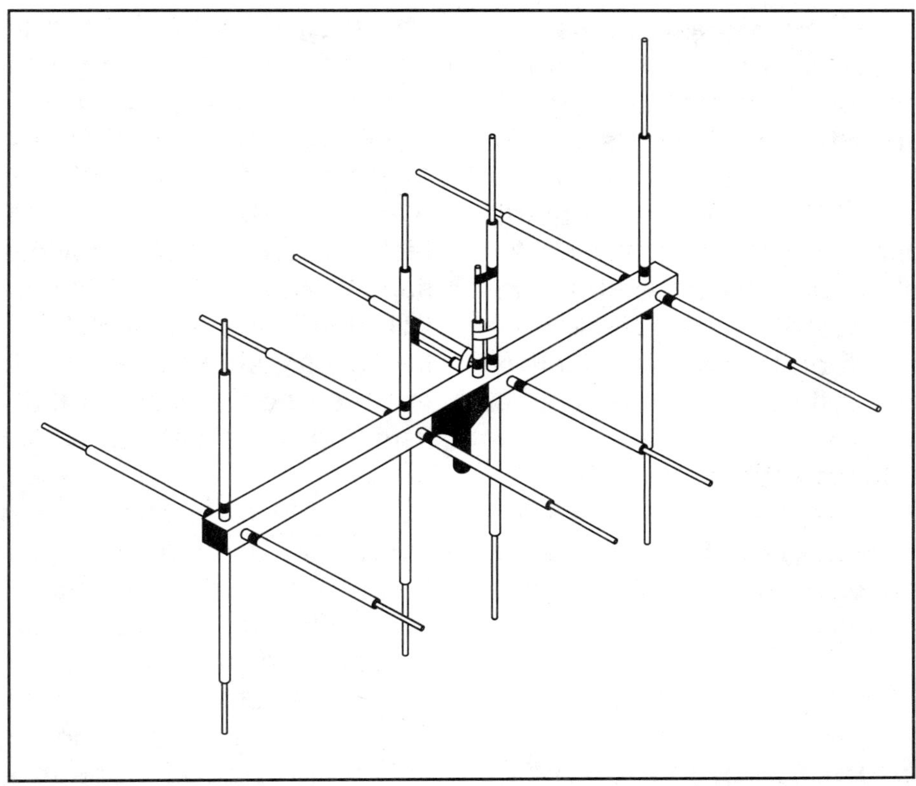

Figure 6-16 *Cross polarized beam antennas like this Jo Gunn JG-4 +4 "out-work" regular beams by taking advantage of polarity differences, as discussed in text.*

on a tall mast, whereas a beam usually requires a tower and rotor for support and pointing.

For mobile operation, a 96 to 101 inch tall stainless steel whip is tops in performance. Unfortunately, it also hits low-hanging tree limbs and requires a sturdy auto body mount. An excellent alternate choice is one of the 6 to 7 foot tall fiberglass/helical wound base and stainless steel "stinger" versions. Shorter 4 foot tall base-loaded antennas are the next step down. For fairly good performance, the 2-foot "minis" and cell phone "look-alikes" represent the minimum in acceptable antennas.

A good base station would be a high-gain vertical like the Valor Pro-Am PS27 shown in Figure 6-14 is an excellent choice. This particular antenna consists of a half wave coil-loaded radiator positioned over a quarter-wave radiator. It stands 18 feet tall. It produces 9.9 dB of gain, which means it delivers an effective radiated power of approximately 40 watts from a regular 5-watt CB transceiver. Assuming one has the time and resources to install a tower and beam-type antenna, it is the ultimate choice for communicating with stations in a selected direction while minimizing interference from stations in other directions. A popular high gain, 8 dB, vertically polarized beam, the Jo Gunn JG-4 is shown in Figure 6-15. Every 3 dB of gain is comparable to doubling a previous effective radiated power level, so the JG-4 makes a 5-watt CB unit sound like 35 watts.

Cross polarized beam antennas are also popular for serious long-range CB communications, because they can be switched between vertical and horizontal plane elements for optimized transmission and reception. Straight line-of-sight CB communications are usually vertically polarized like the CB antennas on autos; however, a distant signal skipping off the earth's ionosphere or reflecting off buildings can change its polarity. The difference in strength of vertically and horizontally polarized signals can be up to 15 dB, and a dual polarity beam antenna like the Jo Gunn JG 4+4 shown in Figure 6-16 takes advantage of that fact. Additional choices in CB antennas are endless. Check with a local CB dealer, study some CB magazines, and select one best suited to your needs.

Power Sources for CB Gear

Modern CB units are fully solid state and operate directly from an auto's 13.8-volt DC battery. They are quite adaptable to alternate/emergency power arrangements. When used in the house, a 12-volt DC (3 or 4 amperes) power supply plugged into a regular 115-volt AC outlet is sufficient for powering the unit. Assuming one connects a pair of heavy-duty clip leads (positive "red" to positive and negative "black" to negative) between a rechargeable 12-volt gel cell or motorcycle battery (or even a non-rechargeable 12-volt lantern battery) and a CB, the unit will work nicely. How long it can be used depends on how much current the battery can supply and how much current the CB unit draws. Current requirements also vary between transmit and receive modes. Check your unit's manual for power requirements.

As an example of estimating operating time, let's assume specifications in your CB unit's manual indicate it draws/uses 0.5 amperes on receive and 2 amperes on transmit. Let's also assume your rechargeable battery is approximately 4 x 3 x 6 inches (H,W,D) and marked as "12 volts, 6 amperes." That means when fully charged, it will supply 6 amperes for 1 hour (6 ampere-hours in technical terms) before losing all power. But suppose you only use 2 amperes. The battery will then last three hours before dying. Ah, but no one is going to transmit continuously for three hours. Most of that time is spent listening, right? At 0.5 amps per hour for receiving, a 12-volt, 6-ampere battery will last 12 hours. Now reduce operating time to 8 hours and visualize receiving 5 hours (0.5 amps per hour x 8 hours = 4 amps) and transmitting for a combined total of 1 hour (2 amps). The battery would then be dead. Substitute current figures for other CB units and batteries into the previous discussion and you have a good working knowledge of battery life. One final tip: Most variety store, non-rechargeable lantern batteries are capable of delivering 6 amps total (remember that means 6 amps for one hour, 2 amps for 3 hours, 1 amp for 6 hours, 0.5 amp for 12 hours, etc.). Keep plenty of fresh batteries on hand (their ampere rating decreases with age), and always be prepared for unexpected emergencies.

Citizens Band radios are widely recognized as a general public-type communications medium when traveling, and they are a good information source during traffic jams. The vast number of people using CB radios can also provide good insights into happenings in nearby areas when one is confined to a home base. Many communities use CB in a quasi-party line manner and in "block watches," giving the medium additional importance worthy of consideration. CB radios are inexpensive, and well worth the investment. Adding a CB radio to your communications capabilities is good thinking.

CHAPTER SEVEN

Personal Communications Systems

Most people will agree that life is an ongoing series of predictable and unpredictable situations: some are good, some not so good. The not so good times are our challenges, and the good times are our rewards. The further this invisible pendulum of life swings from center to side, the greater its variations at each extreme. Being prepared for any and all unusual situations may not be feasible, but knowing all the options at your disposal is always advantageous. In light of these occurrences, this chapter presents an informative overview of several personal communications systems adaptable to a wide range of situations. We will discuss low power business band "talkies," portable marine band equipment, GMRS and FRS handhelds, cell telephones, 49 MHz "free band" units, and field telephones. Armed with knowledge of these short-range units and other monitoring gear discussed in this book, you can feel confident to handle almost any communications need which may occur.

The equipment covered in this chapter operates within the VHF and UHF ranges, so let's briefly review some noteworthy facts of signal propagation in these upper frequency bands. First, stand-alone VHF/UHF equipment (like portable transceivers not dependent on a repeater or relay station to extend their coverage area) are typically limited to a few miles line-of-sight communications range. As such, their frequencies of operation can be "reused" in different areas. This concept usually works out fine; however, we should be aware that unusual weather conditions like temperature inversions occasionally conduct high power (repeater-relayed) VHF/UHF signals 100 to 300 miles. Temperature inversions occur mainly during spring and fall when warm air becomes trapped within layers of cool air. It looks like a layer of dust or smog on a far horizon.

Second, VHF signals get through foliage and wood structures but are restricted by metal objects and buildings. UHF signals are attenuated slightly more by foliage and wood but reflect off metal objects and buildings. VHF equipment has a slight advantage for outdoor/countryside applications; and UHF equipment is slightly more beneficial for use around buildings, offices, and underground parking decks. With these aspects in mind, let's now consider some two-way systems available from communications dealers nationwide.

Handheld Business Band FM Transceivers

Two-way radio operation in the business band or private land-mobile service is used by industries, companies (large and small), and public safety groups for communications among personnel. Users include utility companies, warehouses, construction crews, and more. When licensed, companies are assigned a specific frequency in the 150- to 173-MHz range for VHF communications or a frequency in the 457- to 470-MHz range for UHF communications. An FCC-issued license is required before equipment can be used for transmitting. Extended range repeaters are

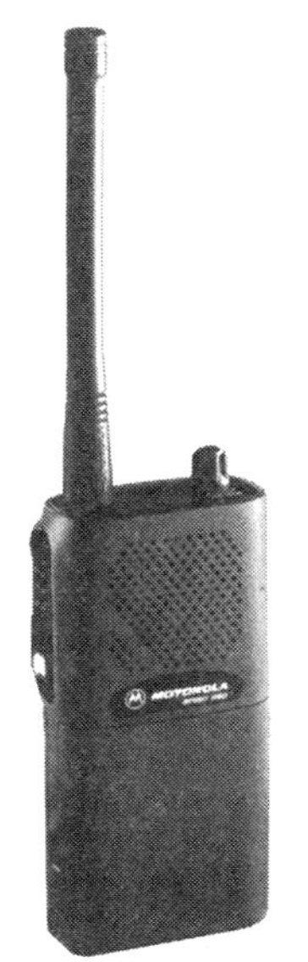

Figure 7-1 Business band talkies require licensing before use, but provide reliable and semiprivate operation over a 2- to 5-mile range. Motorola brand units are famous for their rugged construction, strong performance, and easy operation.

also used in some commercially related applications.

Two frequencies within the VHF business band range, 154.570 MHz and 154.600 MHz, are allocated for personal/business use with handheld transceivers. Such handheld units are available from radio and electronics dealers nationwide (see Figure 7-1). These transceivers are usually supplied with a rubber ducky antenna, rechargeable battery pack, charger, and an application for a license. A new owner should fill out the application, include the listed filing fee, and not transmit with the units until receiving a valid license from the FCC.

Business band FM handhelds are usually rated at one or two watts of power and marked with a blue dot/sticker to indicate operation on 154.570 MHz or with a green dot/sticker to indicate operation on 154.600 MHz. Range is usually 2 to 5 miles, depending on terrain and obstructions. A relatively small number of people use business band handhelds and they must be licensed, thus communications are more "private" than "public" in nature (although anyone with a scanner can tune in business band frequencies). A special channel within the UHF business band, 464.550 MHz is also allocated for use with handheld FM transceivers. The same working concept applied to the VHF handhelds is used here. The only credible difference is that UHF is more suited for use in confined areas such as buildings. Rubber ducky antennas for UHF handhelds are shorter than their counterparts, and their unobtrusive nature may be a consideration in special applications.

Business band talkies are especially designed for easy "no experience necessary" operation and built for strenuous use. Often, a top-mounted volume control and transmit LED (Light Emitting Diode) and a side-mounted PTT (Push-to-Talk) switch are the only exposed controls. If several members of a group use business band handhelds, an optional CTCSS encoder/decoder (Continuous Tone Code Squelch System) may be installed in each handheld and set to recognize a specific subaudible tone. Using this arrangement, several users can share a single frequency, and each user's handheld will respond only to specific calls for that unit.

Marine Band FM Talkies

Are you involved in boating activities near the shore or living close to coastal areas or major waterways? If so, you may find using marine band handheld FM talkies beneficial for short-range communications. As you will recall from our chapter on scanning, a large number of channels in the 156- to 162-MHz range are available for use by ships and small crafts within 10 or 20 miles of shore. The ability to contact other boats and mariners in addition to shore-based stations, marinas, etc., has numerous benefits, especially during emergencies. Further, marine handhelds are equipped to receive NOAA weather broadcasts in the 162-MHz band, a double advantage. Larger and more elaborate VHF marine transceivers are aboard commercial vessels, Coast Guard cutters, and the like. A small handheld marine transceiver is fine for contacting them over a distance of a few miles, and it could prove a real lifesaver during a crisis. A formal band plan specifying which channels to use for general communications, ship-to-shore operations, etc., is employed on the VHF marine band and an FCC-issued license is required for operation. A license application, channel list and band plan, and guidance for getting started is packed with a new marine band handheld or available from marine equipment dealers nationwide. Many shortwave, amateur, and two-way radio dealers also carry a good line of marine radios.

A recent amendment to FCC rules now allows private land-mobile users to share 18 of the VHF marine band frequencies/channels for inland communications. "Inland" is described as being at least 125 miles from an existing navigable waterway or major coastline. The 18 allocated frequencies are the full duplex telephone channels between 157.20 and 162.00 MHz. Although some inland services ignored FCC rules concerning the marine band in the past, and used marine talkies for business or personal applications, this activity bordered on being illegal. Possibly the FCC realized that fact when amending its rulings.

VHF marine handhelds typically run 5 watts of power, with a battery-conserving selection of 1 watt, and may cover many more than the inland-authorized channels. Furthermore, a pair or group of selected units should be capable of simplex operation (same transmit and receive frequency) rather than full duplex operation (separate transmit and receive frequencies) when used in the newly-authorized 157.20- to 162.00-MHz range. If two (or more) handhelds are used full duplex style, each will not be able to receive the other. Finally, remember using inland-authorized (157.20 to 162.00 MHz), talkies around coastal areas is not advisable, as it may unknowingly interfere with (full duplex) marine telephone operations.

GMRS and FRS Units

The General Mobile Radio Service operates on a set group of channels/frequencies in the 462 MHz/UHF range, as shown in Figure 7-2. The service has been established for non-commercial groups needing a reliable means of communication within a 2- to 10-mile area. Users include search and rescue squads, volunteer firemen, and volunteer groups assisting in community events such as parades. Some of the groups utilizing GMRS have access to technical assistance, thus range-extending repeaters similar to those used in Amateur Radio 2-meter and 70-cm activities are operational on a few GMRS channels. Various groups know

Base and Mobile Frequencies	Repeater Input Frequencies
462.550 MHz	467.550 MHz
462.575 MHz	467.575 MHz
462.600 MHz	467.600 MHz
462.625 MHz	467.625 MHz
462.650 MHz	467.650 MHz
462.675 MHz	467.675 MHz
462.700 MHz	467.700 MHz
462.725 MHz	467.725 MHz

Figure 7-2 *Outline of the UHF band frequencies/channels used in GMRS. Note: 462.675 MHz is the emergency/REACT channel.*

Figure 7-3 *Maxon's popular 210+3 GMRS handheld features 10-channel operation, 2- or 5-watts output, scanning, CTCSS calling, and more. The unit is supplied with a rechargeable battery pack, charger, and cable for operating the unit in an auto.*

(or become informed) that communications should be restricted to their specific group, and unauthorized use of another group's repeater without financially supporting that repeater should be avoided.

Soon after the inception of GMRS, the CBers famed REACT emergency/assistance service and several GMRS groups joined forces. The GMRS channel of 462.675 MHz became the "emergency channel." Today, monitoring of 462.675 MHz by REACT volunteers is growing and expanding quite rapidly. Since the UHF/462-MHz range used for GMRS reflects off buildings and metal objects such as autos, it provides good coverage in metropolitan areas which often attenuate CB signals. Further, GMRS is "common ground" that can be used by REACTing CBers and emergency squads alike. In fact, anyone in need of interference-free short range communications for non-business applications can use GMRS quite successfully. An application for the necessary license is packed with each new GMRS transceiver sold. Simply fill it out, include the small filing fee, and mail it to the FCC. A valid license will be issued to you soon thereafter (see Figure 7-3).

Today, strict control over the sale of GMRS handheld talkies has become impossible. The FCC has downsized and lacks the personnel to monitor GMRS frequencies. In addition to the groups and REACT-affiliated services mentioned earlier, users now include auto dealerships, campers, boaters, possibly some home businesses, and others (Figure 7-4). Getting operational is easy: Just step up to the counter at an electronics supplier, plop down the plastic, and walk out with two or more GMRS handhelds. In response to such unrestricted access, the FCC issued a statement (actually a description) emphasizing GMRS was and is not a low-cost alternative for business

Figure 7-4 *The widespread availability of GMRS handhelds make them ideal for search and rescue work and hiking or camping when a group is spread out over a large area.*

Family Radio Service	
Channel	**Frequency**
1	462.5625
2	462.5875
3	462.6125
4	462.6375
5	462.6625
6	462.6875
7	462.7125
8	467.5625
9	467.5875
10	467.6125
11	467.6375
12	467.6625
13	467.6875
14	467.7125

Figure 7-5 *FRS, the Family Radio Service, uses 14 UHF channels located between GMRS channels.*

band-type communications, nor a service like Amateur Radio. Neither was it a "free-for-all" like CB. That makes sense. Businesses prefer more private frequencies and the UHF range for GMRS is not well known by the public at large, so overcrowding is not yet serious.

With GMRS handhelds readily available to everyone and growing in popularity, industry is applying pressure for more GMRS channels. The FCC approved a new Family Radio Service (FRS) in the summer of 1996. Fourteen channels located between regular GMRS channels have been allocated for this low-power medium (Figure 7-5). The FCC defines FRS as "a low power, short range UHF service established to meet the communication needs of families and groups." The term "groups" was left undefined and possibly open to individual interpretations.

Described in a nutshell, the purpose of FRS is to provide a communications medium for families to stay in touch when separated by short distances (e.g., in malls, buildings, living or working in the same neighborhood, etc.) A license is not required.

What is the difference between a GMRS and an FRS handheld? An FRS talkie is usually lower priced, and its power output is less (one-half watt, maximum). An FRS unit also operates on channels between GMRS channels. Operating channels/frequencies can be checked and confirmed by tuning in the unit's signal on a scanner (convenient for determining if unfamiliar units are GMRS or FRS).

At this point, the use of FRS talkies for survivalist applications should be obvious. Family members go in different directions each day, but can instantly communicate independently of telephones or normal power systems with FRS. FRS is ideal for staying in contact with family members at unfamiliar stops while traveling (Figure 7-6). Assuming more REACT groups monitor FRS channels, emergency assistance will be widely available. Carrying an FRS talkie definitely

Figure 7-6 *Ultra compact FRS talkies do not require licensing and give family members an interference-free means of local area communications. Units operate on UHF channels, they are impressive in performance, and could easily prove to be a lifesaver in an emergency.*

beats not having any form of personal communications and it could easily prove to be a lifesaver in a disaster situation. A word to the wise should be sufficient.

Cellular Telephones

These pocket-size wireless telephones are more popular today than CB radios were during their heydays of the 1970's. Most people are familiar with their use, so our discussion will be brief and to the point.

A cell phone is basically a full duplex (simultaneous transmit and receive) transceiver operating in the range of 860 to 893 MHz. It transmits to, and receives from, "cell" sites located throughout every major and minor metropolitan area in the United States. Each cell site has its own transmitter, receiver, and computer system connected to a central telephone switching center that keeps track of all telephone messages going to and from cells by subscribers. As you approach the range limit or fringe area of a cell while talking on the phone, the telephone switching center hands your call off to an adjacent cell. This goes on throughout your travel within a particular area. When you are not using the phone, the computer tracks/remembers the last location of use and can send calls to you when someone dials your number. When you place a call, the computer knows who you are, checks to ensure your account is paid up, then brings up a dial tone and records your outgoing telephone number. Cell sites are installed in all densely populated areas, but are rare in remote areas. Consequently, a cell phone is basically useless in desolate parts of various states, or when you are boating more than 30 miles off shore. Use of a cell phone during a personal emergency can be quite beneficial for summoning help. Its use during a widespread emergency or crisis, however, is a different situation. Cell cites rely on AC power for operation. Some may include a backup battery system, but it would be overtaxed by a hoard of users during a major outage of telephones and power. Hope is on the horizon, however.

Cell phones with the capability of direct satellite access are being developed. As they evolve, their initial cost-per-minute of use is more "astronomical" than "affordable." As this new technology becomes more commonplace, prices should drop. Satellites within range of an emergency/crisis area, however, will still be overtaxed and loaded down when everyone attempts simultaneous use. Cellular phones are super-convenient items that may save lives, but I definitely would want an alternative means of backup communications.

49 MHz "Communicators"

At one time or another, almost everyone needs an inexpensive means of very short-range communications, possibly with "hands free" operation. Some examples are a small group of tourists visiting gift stores, a parent keeping track of a child while shopping (Johnny, where are you?), and motorcyclists traveling together (Figure 7-7). Another example might be using communicators so an invalid grandparent in an adjacent house or added room could call for assistance during the night by simply talking loudly. A pair or trio of small 49 MHz two-way communicators made by companies such as Maxon, Midland, and Radio Shack will address those requirements in a convenient manner.

49 MHz equipment has been in existence for several years. Indeed, units with voice-activated transmit (VOX) operation and all-in-one earphone and mike combinations

__Figure 7-7__ The ability to converse in a flexible and hands-free manner rather than achieving maximum range is often beneficial. Couples motorcycling or bicycling together is a good example. Addressing these needs are voice-operated 49 MHz communicators of various types.

are often used by bouncers in roadhouses. Other equipment using 49 MHz include older model cordless telephones (newer models operate on 900 MHz) and wireless baby room monitors, so interference is possible in crowded areas like apartments and condos. 49 MHz is often referred to as "free band" because this frequency range is used for a mixture of unlicensed and ultra low power applications. The typical range of 49 MHz equipment is 1/8 mile, or one city block, unless the signal is obstructed.

__Figure 7-8__ This 49 MHz communicator has voice activated transmission, earphone and boom mike adaptable to a cycling helmet, a socket for an intercom cable to a passenger, and a 0.25-mile range on one of 5 selectable channels.

49 MHz units with the selection of Push to Talk (PTT) or Voice Operated Transmit (VOX) can prove useful in numerous situations. Cycling is one example; calling for direction while working on a project requiring both hands is another example. Some popular styles of 49 MHz communicators are shown in Figure 7-8, 7-9 and 7-10. The units are small and include belt clips. One model has an earphone and mike designed to fit in a helmet and one model has a combination mike/ear bud (small earphone) for inconspicuous operation. A noteworthy fact to remember about 49 MHz communicators is that they are not professional-grade units and their range is somewhat unpredictable. I would prefer to triple check their range before declaring them fully reliable over a specific distance for disaster communications.

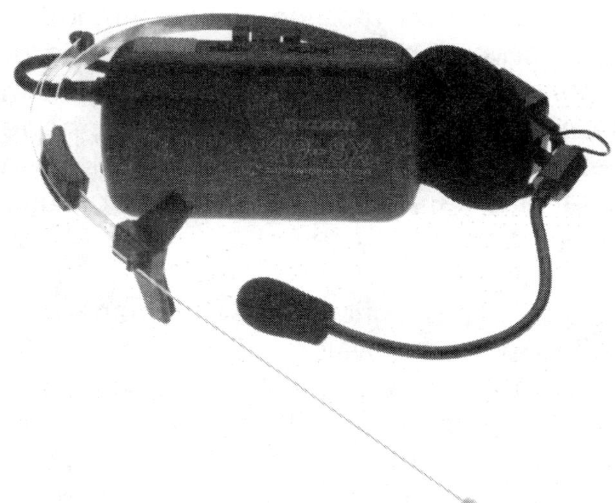

__Figure 7-9__ An inexpensive single channel 49 MHz communicator. The unit holds appeal for partners working in a sound-conscious environment.

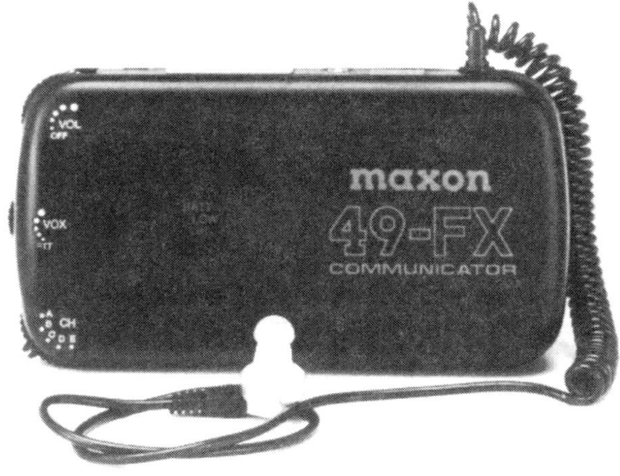

Figure 7-10 The deluxe 49 MHz communicator clips to a belt and the flesh-colored earplug doubles as a mike. The result is an inconspicuous, 2-way unit for use in discrete applications.

Field Phones

Need a private and secure method to talk between buildings, remote cabins or tents? Desire independence from tap-in monitoring by commercial telephone services? Field telephones make an ideal auxiliary and quite secure ground communications system.

Simply described, field phones are a battery-powered system utilized for specific point-to-point communications. They are supplied with a handset and base and are interconnected by a user-supplied wire. A hand crank on the base unit is used for "ringing up" the distant phone. Ringers are adjustable in volume to conform to different situations or applications. The use of a hand crank allows the batteries to stay disconnected and maintain full charge until either unit's handset is lifted from the base. The range of field telephones is more dependent on the amount of wire you can obtain than the phones themselves: A good set of field phones can work up to 20 miles.

Field phones are available from specialized communications equipment dealers and larger military surplus dealers. One of the more impressive units on today's market is the military TA-312/PT system shown in Figure 7-11. It is available through Major Surplus and Survival.

New styles and types of personal communications systems continue evolving every year, and staying informed about the latest technological advancements is strongly encouraged. How? By reading a wide cross section of electronic and communications magazines. Scan them for what you find useful to your lifestyle rather than taking time with "casual reading." Another worthy idea is visiting one of the rapidly growing "spy stores" surfacing in larger cities. There you will find locators and tracking units, short-range "bugs," miniature wireless video cameras, and more. Check your area's telephone directory for dealers. Always use personal communications systems wisely.

Figure 7-11 A pair of surplus military field telephones can provide a two-wire communications system up to 22 miles. The units are powered by 2 "D" cells and provide secure communications. These field telephones can be found in many military surplus stores and surplus mail order houses— effective low-cost communications.

CHAPTER EIGHT

GPS Receivers: Personal Pathfinders

Whether caught in a surprise snowstorm, boating out of sight of land, or camping in unfamiliar territory, the ability to navigate or pinpoint a location without markers is a vital aspect of self-reliant survival. Until quite recently, a compass and a map were the traditional instruments for finding your location and direction. Today, however, a new electronic technology known as the Global Positioning System (GPS) is making navigation on land, on the sea, and in the air highly accurate and accessible to everyone. Indeed, GPS is now being described as one of the prime new utilities for our modern age. Why is this so? Unprotected and unguarded traveling is more unpredictable and more dangerous today than ever before. No one knows whether a person asking directions is sincere or a scoundrel; nor does the person asking directions know if he/she will be assisted or attacked. Self-reliance is the only safe answer.

What is GPS and how do you use it? Simply explained, it is a means for finding your location or position and plotting a path with electronic markers anywhere in the world. Accuracy is within 25 to 50 meters when using a pocket-size and relatively inexpensive GPS receiver, and more elaborate GPS units are coming in the near future. GPS, complete with an automatic position reporting system, is also being integrated into many new automobiles and rental vehicles to assist travelers in navigating unfamiliar areas. A vehicle is never lost, so to speak, as a computer at a main office knows precisely where it is all the time. Equally impressive are the ultra-accurate maps being developed for GPS use. They show more details about an area than even old-timers and surveyors know! Indeed, GPS is now being used for high precision surveying in all world areas and will probably become the industry's standard of reference by the year 2000. On the personal applications side, you can carry a handheld GPS receiver in a coat pocket or place it on a vehicle's dash

Figure 8-1 *Pocket-size GPS receivers are the ultimate personal navigation aid. They read out latitude and longitude accurately, indicate bearings, routes, return paths, and much more. The unit shown here is displaying a return path for a route stored in its memory.*

and it will determine and read out your location in latitude, longitude, degrees, and minutes from almost anywhere (See Figure 8-1). When traveling, you can also enter waypoints along the route and the GPS unit can reverse them and plot a return path. It is a superb location and direction-finding utility.

How GPS Works

As illustrated in Figure 8-2, a constellation of 24 uniquely designed satellites comprise the Global Positioning System network. Twenty-one of the satellites are in continuous operation and three are spares, or "backups." This number ensures a minimum of four satellites are always within line-of-sight range from any point on earth and it also provides immediate replacement should any of the satellites develop problems. The satellites are located 11,000 miles above the earth, each in an orbit of progressive inclination with reference to the equator. The satellites were developed by the U.S. Department of Defense and are constantly monitored by them. Their main function is to serve as a locating/positioning system for the military (when used with the military's high budget equipment, the GPS can pinpoint locations to within three millimeters). For civilian use, however, commercially available handheld GPS receivers enable anyone to use the GPS satellites for highly accurate position locating and direction-finding assistance.

A handheld GPS receiver looks like a relatively simple device, but it is actually a marvel of modern electronic technology, and possibly the most advanced item in its price class available to the public. A GPS receiver contains a very sensitive spread-spectrum microwave receiver tuned to 1575.42 MHz and an extremely accurate clock for measuring travel time or radio signals (which move at the speed of light: 186,000 miles per second). It also contains a quite sophisticated mathematical unit, which performs trigonometric and algebraic calculations on measurements to determine locations, routes, etc. The receiver uses signals from GPS satellites as reference points for triangulating its exact point on earth. Measurements from three satellites provide triangulation data, while coded data information from a fourth satellite corrects for timing/measurement errors. The previous description of operation was highly simplified for ease of understanding, but calculations within a GPS receiver are actually quite complex. Measuring distances from three orbiting satellites by radio signals, for example, involves calculating time in nanoseconds, one billionth of a second! Each GPS satellite uses an expensive atomic clock, continuously checked and corrected in its data-stream

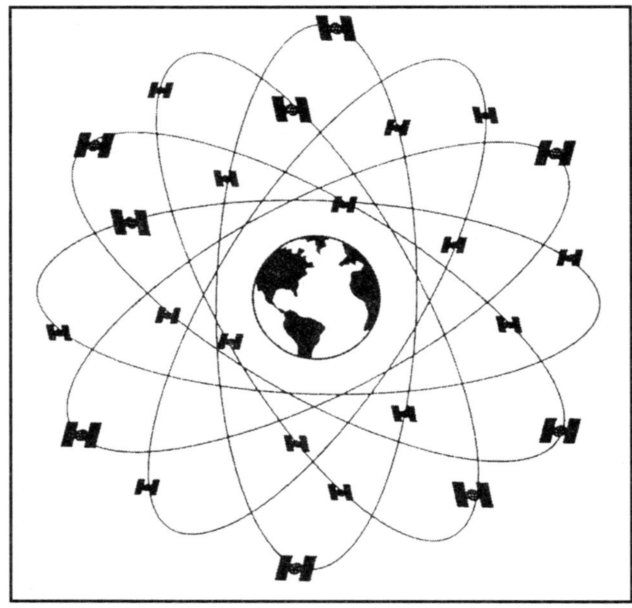

Figure 8-2 *A constellation of 24 satellites orbiting the earth comprises the GPS network. Each satellite transmits precise time and triangulation data used by GPS receivers for pinpointing one's exact location anywhere in the world.*

transmissions, to maintain incredible accuracy. Handheld GPS units do not contain an atomic clock, which typically costs $100,000 each, so they use a fourth satellite's transmitted data as an "accuracy-assuring substitute." Since a highly sophisticated calculator is integrated with the GPS receiver, it also serves other functions such as plotting paths, indicating directions, etc.

Earlier, we mentioned all GPS receivers are tuned to 1575.42 MHz, the transmission frequency of all GPS satellites. To put that in perspective, this frequency is above the cellular telephone band (860-890 MHz, approximately), and below the popular Cable/"C" band used in home satellite TV reception systems (3,700-4,200 MHz). The frequency of 1575 MHz or 1.56 GHz is strictly line-of-sight in nature, which means a GPS receiver's antenna must have a clear sky view to find/receive signals from GPS satellites. The tiny antenna is usually inside a small enclosure near the top of the GPS receiver. The antenna can be removed and mounted remotely on some receivers, a real advantage when traveling in enclosed vehicles. That is because the antenna must electrically "see" four satellites at the same time and from the same position to acquire triangulation data. Considering all the stipulations and constraints, one might assume inexpensive commercially available GPS receivers are slightly limited in performance. Such is definitely not the case! GPS receivers put all other forms of direction-finding equipment out to pasture. They are totally perfected, here to stay, great investments in self-reliance, and available from communications dealers and marine equipment suppliers nationwide. Check one out, it could prove to be a genuine lifesaver when you are lost in strange territory and need fast, accurate position location.

GPS Maps and Mapping

Equally impressive in the area of GPS technology are the snap-in mapping database cards for handheld receivers and the software so laptop computers may be used with GPS receivers. Some of the snap-in cards show several states and major roads, rivers, airports, etc. Other cards get more exact on specific areas and display individual streets in metropolitan areas with high accuracy. Additionally, your present location, speed, route of travel, etc. can be plotted on the display so the GPS unit becomes a full travel atlas!

In laptop software, the mapping capabilities are even more extensive. Displayed areas can be zoomed in and out, colors can be added, and even the smallest streets located with ease. Electrically, the laptop must be connected to a GPS receiver via an NMEA port for this arrangement to work. Many, but not all, handheld GPS receivers include an NMEA port. The area of GPS software for computers is growing at a phenomenal rate. Check with your favorite communications equipment/GPS dealer for the latest news in new GPS software.

Previewing GPS Receivers

As additional familiarization with GPS, let's briefly preview some of the popular handheld receivers. This preview will give a good idea of their similarities and differences, and serve as an introductory reference point from which to visualize the type of unit that would best fit your own needs and lifestyle.

First, all of today's GPS receivers require initializing, or basic start-up programming when first used or after being left unused without battery power for long periods. This is because the receiver does not know where it is and must be referenced on at least four satellites to "get its bearings." Some receivers are supplied with initializing data for cities within a 200 or 300 mile range that can be entered manually to reduce "start-up" time, a most convenient asset. Finally, there are considerations of size, weight, and cost that ultimately influence which receiver is right for you. Let's look at some well-known GPS receivers.

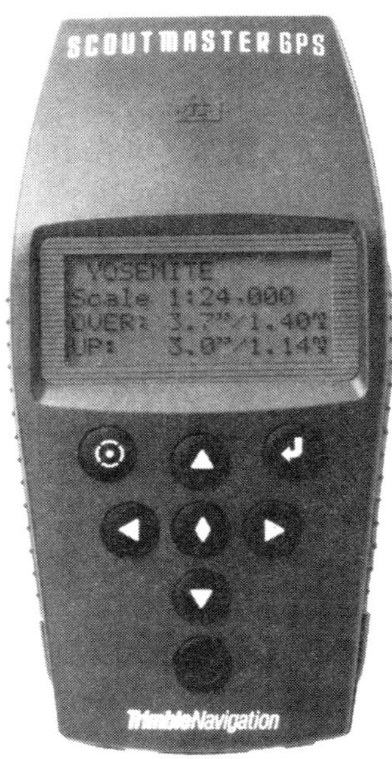

Figure 8-3 *The ScoutMaster GPS receiver produced by Trimble Navigation displays its location in latitude and longitude, plus time, distance, and direction to a desired destination. Unit also has multiple displays, including a "measure over and up" feature (shown) for marking reference points on topographical maps.*

The Trimble ScoutMaster One of the most popular GPS receivers today is the ScoutMaster made by Trimble Navigation Limited. This unit, shown in Figure 8-3, measures 6.8 by 3.3 by 1.3 inches (H,W,D), and operates 6-10 hours using 4 "AA" batteries. It has a high-speed mode for fast calculations, will track up to eight satellites, and displays latitude and longitude with accuracy to parts of a second. The ScoutMaster will also display information in the Universal Transverse Mercator (UTM) mode, which can be applied directly to any USGS topographic map. Additionally, the unit includes a unique "Over and Up" mode that allows a user to call up location information in inches "over (to the left) and up (vertically)" from any selected map reference point. Other features include call-up displays for distance and direction to a desired destination, route plotting, route reversing for backtracking, and a 250 location/waypoint memory. The ScoutMaster also has an NMEA port, permitting interconnection to a laptop computer for unsurpassed in-field use. Learning to use the ScoutMaster takes several hours and its display is small compared with other units, but the ScoutMaster's multiple navigating assets and modes truly overshadow these minor drawbacks.

The Garmin GPS-38 Targeting the budget conscious and non-technical minded newcomer, Garmin International produces the model GPS-38 shown in Figure 8-4. The unit features easy initialization by the quick selection of a country and state, a moving map showing information pertinent to a particular trip or route, a rotating compass and bearings display, and a route backtracking display page. The GPS-38 will track up to eight satellites simultaneously, display latitude and longitude down to fractions of a second, and

Figure 8-4 *The Garmin GPS 38 receiver indicates latitude and longitude plus additional call up displays that include a moving map, rotating compass (shown here), and more.*

store up to 20 routes with up to 30 waypoints each. The unit is extremely user friendly, super small (6 by 2 by 1.2 inches H,W,D), water resistant, and equipped with an NMEA port for laptop computer interfacing. The unit will operate 10-20 hours using 4 "AA" batteries. A large multi-page display and easy operation combined with economical cost make the GPS-38 quite attractive.

The Magellan GPS 2000 This high profile receiver is made by Magellan Systems Corporation and distributed through many well-known marine and electronic outlets nationwide (Figure 8-5). The GPS 2000 features four selectable display screens that include latitude and longitude down to parts of a second, bearings and directions, and path plotting and retracing. It can store up to 100 waypoints per route, reverse the route for backtracking, and also display data in UTM for referencing on a topographic map. For easy initialization, the GPS 2000's manual includes quick-entry data for cities and areas around the world. Enter some basic coordinates, set the GPS 2000 in a clear area for a few minutes, and it is ready for use. The unit measures 6.6 by 2.3 by 1.3 inches (H,W,D) and operates up to 17 hours on 4 "AA" batteries. An NMEA socket is not included on the GPS 2000, as it is Magellan's "low end" unit; one of many in an extensive line of Magellan GPS receivers.

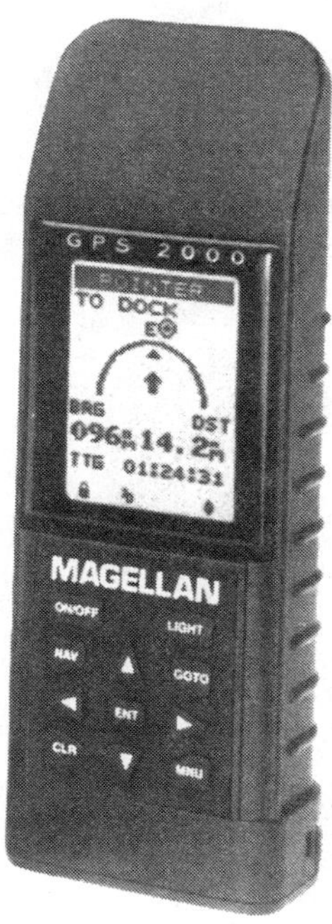

Figure 8-5 *Magellan's Model GPS 2000 features four navigation screens, including the quasi-radar display shown here. It is a beneficial unit for determining position, direction, speed, and progress on land or sea.*

A wide selection of additional brands and models of GPS receivers are available through marine and communications equipment dealers nationwide, and previewing all of them here is not possible. Hopefully, however, we have opened another door in the area of survival communications and expanded your knowledge of various items available on today's market. The GPS can be a tremendous asset to travelers and backwoods pioneers, and we encourage you to further investigate this rapidly expanding area of modern technology. For survival communications equipment, the GPS receiver is a must.

CHAPTER NINE

Free-Playing Radios

In continuing our study of survival communications and their application to numerous situations, this chapter presents details of radios that operate completely independently of normal energy sources. The radios featured in this section do not require batteries or household power for operation, yet they are capable of surprisingly good reception for any length of time necessary.

What types of radios operate without batteries or external power? Classic crystal sets you can quickly assemble at home or in the field are one example. Wind-up radios that can be purchased ready to use from shortwave dealers and survival stores nationwide are another example.

When might such a free-playing radio be beneficial? During exceptionally long power outages or when your usual supply of batteries is exhausted are two obvious examples. Slim chance of that happening, you say? Think again. While writing this chapter, reports of an unexplained power outage affecting several states in the Midwest for almost a day were being investigated. Opinions regarding similar widespread power outages affecting the Eastern U.S. for several consecutive days were also being aired on shortwave talk programs. With government agencies and large computers able to control the distribution of electrical energy today, long-term outages are indeed possible. If any type of widespread emergency occurred (e.g., a mass power outage), Civil Defense affiliated stations using alternate energy systems are set up to transmit welfare and survival information on the AM broadcast band. A self-powered radio can receive that information on a non-stop basis.

When connected to a simple wire antenna, crystal sets are capable of receiving AM broadcast band stations around the country. They can also be modified to receive some of the higher power international shortwave broadcast stations and even local airport control towers and aircraft. Crystal sets are not overly sensitive or selective (i.e., they pick up the strongest stations in a particular range and "fine tuning" is not always possible), but these deficiencies often prove advantageous. You can build a crystal set from spare parts mounted loosely on a pine board, or assemble a working unit on a stained decoupage plaque as a replica of an old-time radio. Make construction a family project, and it can also introduce your children to survival communications. In fact, the possibilities, opportunities, and expansions sparked from initially assembling a first crystal set are endless. What is involved in "home brewing" a crystal radio? Read on!

Components in Crystal Set Receivers

A crystal radio is made up of a coil of wire, a tuning capacitor, a crystal diode, and an earphone. The number of turns of wire comprising the coil determine the set's approximate frequency range. The capacitor is used to select and tune stations within that frequency range. The crystal diode "detects" radio signals, or converts them from radio

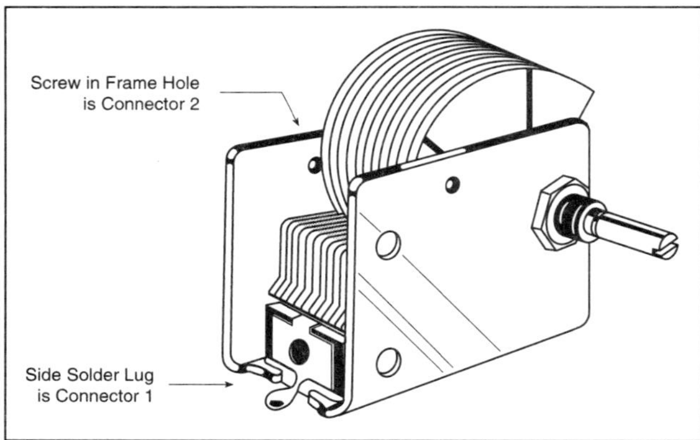

Figure 9-1 *A single section variable capacitor contains a set of fixed position plates and a set of movable plates that mesh without touching.*

waves into tiny voltage variations. The earphone then converts the tiny voltages into sound waves you can hear.

A wire antenna for receiving radio waves also must be connected to a crystal set. Usually, a single wire stretched out 60 to 80 feet and erected as high as feasible makes a good antenna for a crystal set. The antenna wire only needs to be heavy enough to support its own weight, but it may be smaller if it is laid on a roof or hidden in an attic. The wire can be insulated doorbell wire or enamel-coated copper wire. In fact, even usually overlooked objects such as metal rain gutters, balcony rails, or chain link fences can often be used as impromptu antennas for use with crystal sets. How do you determine if such items are usable impromptu antennas? Easy! Try them with your crystal set and see. One stipulation: **NEVER USE POWER LINES AS ANTENNAS! NEVER ERECT AN ANTENNA CLOSE ENOUGH THAT IT (OR YOU) COULD FALL ON POWER LINES! ALWAYS STAY CLEAR OF POWER LINES. YOU COULD BE ELECTROCUTED. THAT FACT APPLIES TO ALL ANTENNAS!** Now let's continue with more notes on the components used to make crystal sets.

Hand-wound coils of all shapes and sizes have been used successfully in crystal radios. Your author has even used thin enamel-coated copper wire wound on a Popsicle stick to make a pocket radio; however, larger diameter coil forms are better for AM broadcast bands and shortwave reception. Some readily available household items that make good coil forms are bathroom tissue or paper towel rollers, any cylindrical plastic or cardboard container, and the ever-famous Quaker Oats® box. Cardboard rollers approximately 1½ inches in diameter are the easiest with which to work. Plastic containers of a larger diameter are slippery but better. The Quaker Oats® boxes (approximately 4 inches in diameter) produce the best and most selective coils. The most popular wire for coils is enamel-coated copper wire of any gauge between number 16 and number 24. Enamel-coated wire has a maroon or wine color. Alternately, similar gauge solid and insulated "hookup" wire works fine. Assuming your selected coil form is large enough to support 40 to 60 turns of wire, regular insulated solid doorbell wire obtained from a hardware store can be used to make the coil. Punch two tiny holes beside each other and near one end of your selected form, then slowly and accurately thread the wire twice through the holes to hold

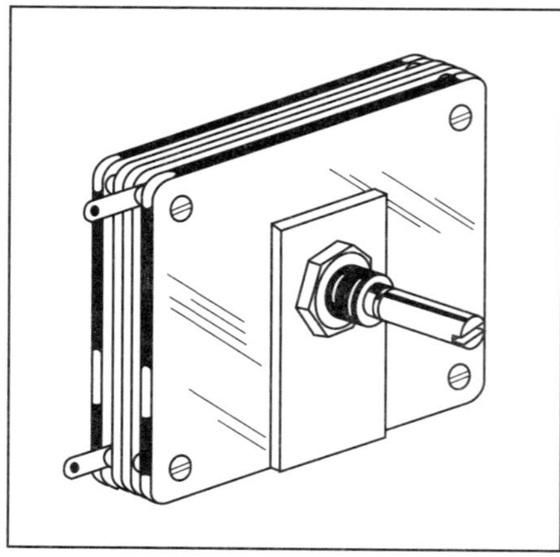

Figure 9-2 *A small wafer-type variable capacitor is used in many portable AM/FM radios and can be salvaged for use in crystal sets.*

one end secure. Be sure to leave approximately 6 inches of wire extended for circuit connections; then slowly and accurately wind 40 to 60 adjacent and evenly spaced turns of wire onto the form. Your first coil may be a bit sloppy, but you can always wind a second and neater coil. Practice makes perfect. When you finish winding the turns, punch two more tiny holes near the form's opposite end. Allow an extra 12 inches of wire to extend at this end for "hookup." Cut the wire and wrap it twice through the tiny holes as before. Finally, remember to remove the insulation or coating from both ends of the coil wire to ensure a good electrical connection during assembly. Scrape off all of the insulation around the ends of the wire with sandpaper or a pocket knife, not just on one or two sides. The uncovered copper will shine like a new penny and serve as a good electrical conductor.

Constructing a tuning capacitor at home is a challenging process, so I suggest purchasing a new one from a local electronics parts store, or better yet, salvaging one from a discarded AM-band radio. The tuning capacitor should be a single-section variable capacitor similar to the type shown in Figure 9-1. Alternately, a thin "wafer type" capacitor used in an AM/FM radio can be used (Figure 9-2). The two connection terminals on a "wafer capacitor" are usually obvious, but finding the two connections on older open-frame variable capacitor requires some study. This is because the capacitor's movable plates connect (through bearings) to the capacitor's frame which is typically connected to the radio's chassis/ground by a mounting screw—usually on the capacitor's bottom. Other holes on the frame's rear can be fitted with a short bolt and nut to secure one hookup wire to the capacitor. The other hookup wire will connect to a side tab that connects to the capacitor's fixed plates. Check your capacitor with a magnifying glass. None of the fixed or movable plates should ever touch each other. If the plates touch, the capacitor will short-circuit the coil and your radio will not work. Also, double check the clearance of your frame connection to ensure that the added bolt and nut do not short-circuit the capacitor's plates. Some capacitors are double or triple section units. Connect one wire to a frame bolt and another wire to only <u>one</u> of the two (or three) side connections. Ignore the other side connections.

Figure 9-3 *A crystal detector with a cat's whisker, ball and rod assembly, holder for galena crystal, and old-style binding posts mounted on a wooden base.*

A galena rock or a piece of Carborundum can be used as a crystal diode, but it is simpler to purchase a diode from an electronics supplier for less than a dollar. We want a germanium diode, type 1N34, 1N60, or 1N82A. The 1N34 is the most popular diode used for crystal sets. A 1N914 computer-type diode <u>is</u> <u>not</u> an acceptable substitute.

The earphones should have a 2000 ohms (or higher) impedance rating; do not use a modern 8-ohm or stereo earphone. "Crystal" earphones are usually 4000-ohms impedance and work fine in crystal radios. If you cannot find earphones (or other parts) locally, contact Antique Electronic Supply Co. In addition to earphones, they also carry coil forms, coil wire, tuning capacitors, crystals, diodes, hook-up wire, Fahnestock connector clips, wire antenna kits and even full crystal radio kits - all at fair and square prices. Since related parts have been introduced, let's now focus on building a "first crystal set."

A Home-Assembled AM-Band Receiver

Assuming you have read our previous discussion of components; secured said items; and have a few hardware items like screws, angle brackets, a pocket knife, screwdriver, and maybe a small soldering iron handy, you are ready to build a free-playing radio. Several variations in the radio's baseboard, coil form, wire, and tuning capacitor are possible. We will begin with a general list of materials needed.

1) Approximately 75 feet of wire for winding the coil. The wire may be enamel-coated copper wire (16 to 24 gauge) or solid wire with plastic insulation (like doorbell wire). Avoid using stranded wire for the coil.

2) A coil form. An empty Quaker Oats® box is ideal. Smaller diameter forms are equally acceptable. Avoid forms less than one-inch in diameter.

3) A germanium (crystal) diode: 1N34, 1N60 or 1N82A.

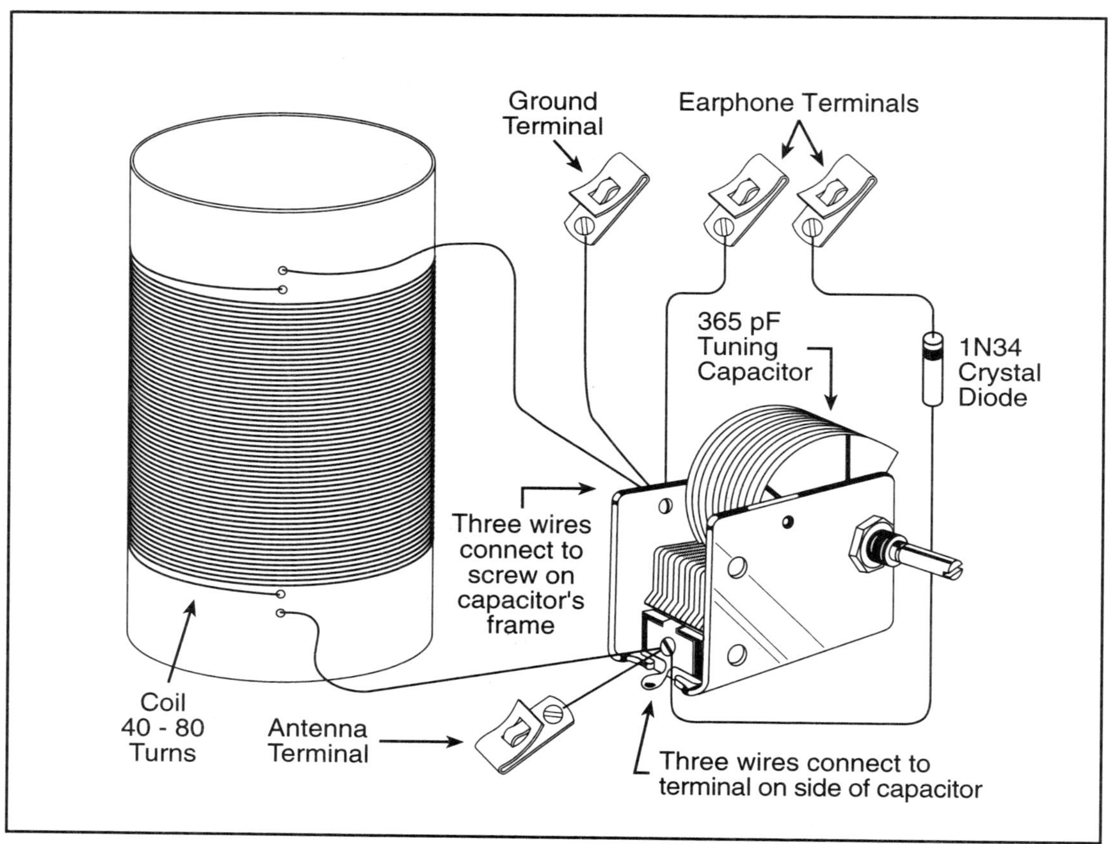

Figure 9-4 *"Open air" view or "pictorial" diagram of the crystal set described in text. Use this diagram as a wiring guide for home assembly.*

4) A 2000 ohm (or higher) earphone or pair of old-time earphones.

5) A 365 pfd. or 500 pfd. single-section tuning capacitor (new wafer type okay; older open-air type ideal).

6) Four Fahnestock clips for connecting the antenna, ground, and the earphone's two wires.

7) Miscellaneous hardware: screws, angle brackets, a knob, screwdriver, needle-nose pliers, small soldering iron, small solder, diagonal cutters, and a baseboard approximately 8 x 6 x 0.5 inches.

8) 100 feet of wire and a pair of glass insulators to make the antenna.

Note: A few pieces of excess coil wire can be used for interconnections. Also, screws with extra nuts can be used for connecting posts in lieu of Fahnestock clips.

Begin by winding the coil and securing it to the form. If you use a Quaker Oats® box, wind approximately 40 turns for the coil. If you use a medium-diameter form (2 to 2.5 inches), wind approximately 60 turns. If you use a small diameter form, wind 60 turns. Remember to leave 6 to 12 inches of wire at each end of the coil for trimming to size and "hookup." You may find it helpful to add a few strips of masking tape to hold coil turns neatly in place, one beside the other, as you go. After completing the coil and securing ends through tiny loop holes, add a few drops of model airplane glue to hold the wire firmly in place on the form. After the glue dries, you can

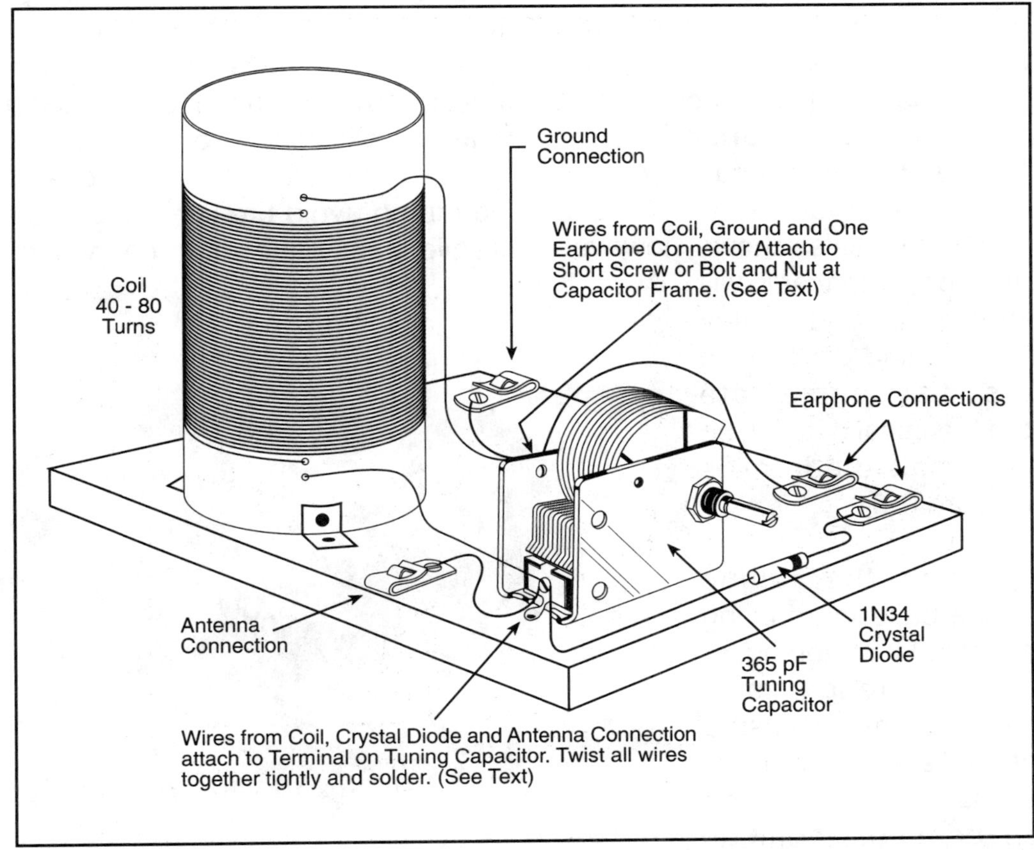

Figure 9-5 *Sketch showing a suggested parts layout for the home-assembled crystal set. Note coil is neatly wound with adjacent turns, and insulation is removed from wire ends before hookup.*

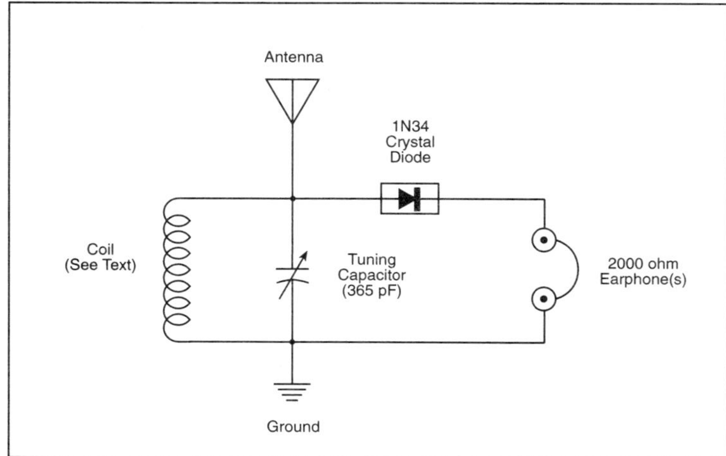

Figure 9-6 *Electronic/schematic diagram of the home-assembled crystal set. Wiring of radio Figure 9-3, 9-4 and 9-5 are exactly the same.*

remove the strips of masking tape. At this point, you can also install small angle brackets on the coil form for stability.

Before mounting the coil on the baseboard, study Figures 9-4 and 9-5 for overall layout guidance. The exact location and position (upright or horizontal coil mounting, for example) is not critical and can be altered to suit your preference. Wiring or "hookup" must be exactly as shown, however. Otherwise, the completed radio will not work. Some of our readers may be familiar with basic electronics (and others may appreciate learning more technical details), so the radio's circuit diagram is included in Figure 9-6.

Next, mount the tuning capacitor on the baseboard using small angle brackets, strapping, or screws inserted through holes drilled in the baseboard. Measure and trim the coil's wires, remove one inch of insulation from their ends, then connect them as follows. Connect one coil wire to a jumper wire routed to the ground terminal and another jumper wire routed to one of the earphone terminals. Then connect the trio to a screw on the tuning capacitor's frame (Figures 9-4 and 9-5). Soldering the wires together is advisable. Alternately, twist them tightly together using needle-nose pliers. Repeat the previous procedure with the coil wire's other end, add a jumper to the antenna terminal, and one wire to the crystal diode. Some capacitors may have a side-mounted screw: Do not use it for a connection, it is a trimmer adjustment. Look closely at the capacitor. Find its side solder terminal lug and connect the trio of wires to that lug.

Jumper wires routed to the earphone, ground, and antenna terminals can be attached with solder lugs or slipped under screws holding the terminals to the baseboard. Screws with one set of nuts for holding jumpers down and a second set of nuts for external wire connections may be used if Fahnestock clips are not available. Be careful to avoid breaking the crystal diode's fragile glass case when connecting it to the

Figure 9-7 *Pioneer Basic Crystal Set. A great beginner crystal set for the kids or the first-time crystal set builder. Tunes well and is a good receiver.*

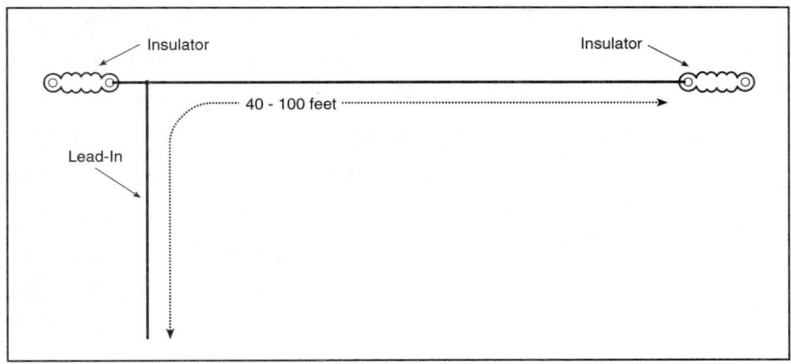

Figure 9-8 *A single end-fed wire antenna 40 to 100 feet long, as illustrated here, works fine with a home-assembled crystal set.*

earphone terminal and capacitor terminal. Diodes are also sensitive to excess heat. If you solder connections, hold each wire between the soldering point and the diode with your needle-nose pliers to dissipate heat before it reaches the diode.

As discussed earlier, a 2000-ohm (or greater) earphone must be used with a crystal set. If your earphone's cable terminates in a single (coaxial) plug, cut it off and carefully remove the insulation to expose its two wires. Connect each one of the wires to each of the radio's earphone terminals. Only one wire to each terminal. Do not allow both wires to touch.

Next, string up an antenna and install a ground connection. As mentioned earlier, a single wire 40 feet long or longer is fine. Figure 9-8 outlines a simple, yet effective, antenna. Stranded wire is recommended here, as it will flex in the wind without breaking. You can secure a good earth ground connection by clamping a wire directly to an outdoor cold water pipe. Do not connect the ground wire to gas or hot water pipes. Only a cold water pipe is earth-grounded. Use your pocket knife to scrape away dirt and produce a shiny, clean connection spot; then attach the wire using a small radiator hose clamp. At this point, you may also wish to add a weatherproof plug and socket assembly so you can unplug the antenna's lead-in and connect it to the earth ground when thunderstorms are expected or when the radio is not in use. Although not a proven fact, there are strong indications that a grounded antenna deters build up of static charges and minimizes the possibility of an immediate area lightning strike.

Route the antenna and ground wires to the radio, being sure they do not touch each other and short (using insulated wire for the ground and antenna avoids shorts), and connect the wires to the radio. Assuming earphone wires are also connected, and you wired the radio correctly, you should now hear sounds (stations!) on the earphones. Add a tuning knob to the variable capacitor and peak the loudest station. Do not expect super sensitivity and ear-bursting volume, unless you are within a half-mile range of an AM station. Also, do not be surprised if one station can be heard faintly when listening to another station. That is a natural characteristic of a crystal set.

After using the crystal set for some time, you will notice a difference in its range and volume during day and night hours. This is because stronger local stations are received during the day. Some of these stations sign off at dusk, and weaker distant stations begin "skipping" into your area. How well your crystal set receives distant stations will be influenced by its antenna's location and height and the coil's dimensions. If you wish to experiment for more sensitivity and selectivity, try this time-proven modification. Using plastic-insulated wire, wind a "primary coupling coil" directly on top, and right in the middle of, the crystal set's main coil. Approximately 20 turns is suggested. Remove

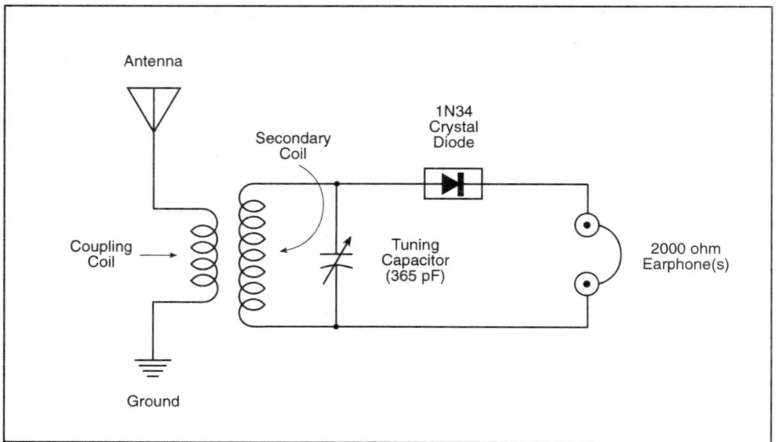

Figure 9-9 *Electronic/schematic diagram of the modified crystal set discussed in text. Addition of a "primary coupling coil" improves reception.*

one-half to one-inch of insulation from each coil end. Remove the previously installed wires routed from the tuning capacitor to the antenna and ground connections. Connect the two wires from your newly installed "primary coupling coil" in their place. Your radio's circuit has now been modified as shown in the diagram of Figure 9-9, and it will probably exhibit improved reception. Congratulate yourself on building a radio from scratch and remember all the assembly details so you can build it again from memory anytime and anywhere needed

Shortwave Reception with a Crystal Set

Earlier in this chapter, I mentioned using a crystal set to receive stronger and higher power shortwave broadcast stations. Since you now understand the basics of crystal set construction, let's consider how it may be modified for tuning from approximately 3 to 12 MHz. Simply explained, the receiver's tuning range is determined by the coil's physical size and number of turns, in addition to the length of the antenna. Reducing the number of turns and shortening the antenna raises the receiver's tuning range. If a Quaker Oats® box form was used, reduce the coil turns to 9. If a 2 or 2.5 inch form was used, reduce the coil turns to 12. Do not use smaller diameter forms. Next, shorten the antenna's length to 20-30 feet (to receive shorter wavelength signals, naturally!). If you prefer using a "primary coupling coil" such as the one included on our modified AM band receiver (a good idea!), wind 6 turns of insulated wire over the center of the previously described shortwave-band coil. That completes the electrical details. Now let's add some operating notes.

First, understand that our crystal set is powered solely by radio energy from the station it tunes in. Strong stations will be

Figure 9-10 *Shortwave Crystal Radio. Tune in the worldwide signals from 3.0 - 14 MHz on a crystal radio. This set is surprisingly simple to build and the reception will amaze you with the proper antenna. Requires no power source. A good little crystal set when the power grid goes down.*

louder than weak stations, and very weak stations will be inaudible. Also remember, the lower frequency shortwave bands are mainly active at night. Do not expect to hear these stations during the day. Some of the higher power stations you may hear at night include WWCR in Tennessee, WEWN in Alabama, The Voice of America, and maybe the British Broadcasting Corporation. Other stations may also be received depending on "skip" conditions, your location, and the hour of the night. Good luck and good listening! Very nice crystal set kits are available from Antique Electronics, (602) 820-5411.

Aircraft Band Reception

What's that? Monitor transmissions from an airport control tower or an airplane on a simple crystal set? Such is indeed possible, but there is a stipulation. These stations use low power, so we must be in close proximity to them for reception. Exactly how close is difficult to define, but 500 feet to a quarter-mile is a good estimation. A word of caution may be advisable, however, as people with strange looking electronic gadgets around airports may be scrutinized by security personnel. On the more positive side, the low selectivity and wide frequency range of a crystal set for the approximate 100-140 MHz range minimizes tuning adjustments. You just get near the source you wish to monitor and listen. Remember, transmissions are not continuous and you may need to listen for several minutes before hearing brief calls.

Rather than modifying our existing crystal set for VHF reception, it is more logical to assemble another set using the same circuit diagram/wiring arrangement as previously outlined, but with more specialized components. Remember the Popsicle stick coil I mentioned in the first part of this chapter? It is a good choice here. Wind approximately 30 turns of wire on the stick, omit using a tuning capacitor, and use a 1N82A or 1N21 diode, and the radio is ready for use. Connect your 2000-ohm earphone, change the antenna to a dangling 3- to 6-foot wire, then try it near an airport. Results can be quite good!

Field-Assembled Crystal Sets

During World War II, soldiers in foxholes often built crystal radios to keep up with the news of the day. Several of their improvising techniques still prove useful today. Lacking a tuning capacitor, for example, they used sandpaper or a knife to scrape insulation from the top surface area of a coil. A sliding wire or piece of coat hanger with its enamel scraped off was then connected to the ground side of the coil and moved along the coil wire's exposed top area for tuning. As the wire or coat hanger was moved toward the "antenna end" of the coil, it would tune and peak stations higher in frequency. As the wire or coat hanger was moved toward the coil's "ground end," it tuned/peaked stations lower in frequency.

Lacking a crystal diode, a single-edge blue blade razor blade and a piece of lead from a wooden pencil were used. Authentic blue blades had a unique metallic property that allowed detection. Stainless steel blades did not (and still do not) work as detectors. Today, only one company makes single-edge blue blades. They are called "Pal Super Single Edge," and are made by The American Safety Razor Co. in Virginia. Alternately, a non-stainless steel paint scraping blade with perhaps a touch of rust can be tried as a substitute. BE CAREFUL! These blades are **sharp** and rust can support tetanus!

Use screws or thumbtacks to secure the razor blade flat onto the baseboard. Next, fabricate a "cat whisker" using a piece of pencil lead held on the point of a safety pin by wraps of wire. Bend the pin's head 90 degrees, insert a screw or thumbtack through it and into the baseboard, then position the lead end so it lightly touches the razor's surface. While listening on the earphone, slowly move the lead's position until the best reception volume is found. Crystal radios are basic, yet quite versatile items, and their applications are truly endless. Remember our previous discussions of them for future reference if and when they are needed!

Figure 9-11 *The BayGen wind-up radio made in South Africa and sold through U.S. outlets. Might this be the first of many modern free-play battery-less radios?*

Wind-Up Radio

Before concluding this chapter on free playing radios, we would like to quickly share details on a recently announced wind-up radio being produced in South Africa. The BayGen unit is shown in Figure 9-11. This portable radio tunes and receives the standard AM broadcast band, the FM band, and the shortwave range of 3 to 12 MHz. The BayGen radio is unique in the respect that it is powered by a small internal electric generator which, in turn, is driven by a wind-up music box type assembly (also built in).

A fold-away hand crank is located on the radio's right side. You wind the crank for approximately 20 seconds (50 or 60 turns), and the radio will play for approximately 40 minutes. Reception is reasonably good and typical for lower cost units. In other words, it is not stereo or high fidelity. It is a rugged little utility item designed to take abuse and tune various stations in an "always ready" manner.

The BayGen was originally designed for use in South Africa and other developing areas of the world where batteries are expensive or non-existent and the need for knowledge is paramount. Considering its significant appeal in survival communications situations, it is highly possible that both the BayGen and units of a similar concept will evolve in the near future and will become quite popular. The BayGen radio, incidentally, is presently available through various dealers.

Some people may visualize crystal sets and wind-up radios as "low tech" and of minor significance, but such is definitely not the case. Everything in our world of electronics serves a particular purpose, and knowing how to apply this equipment to various situations is a priceless asset. Knowledge is power.

CHAPTER TEN

Weather Information Sources and Resources

The most well-known and often experienced emergencies or anxious situations affecting all of us are related to weather disturbances like snowstorms, floods, hurricanes and tornadoes. No locale is immune to weather variations. Even the most seemingly perfect areas are susceptible to electrical storms, excessive winds, or extremes of temperature. And while we may not be able to change the wrath of mother nature, being accurately informed of impending circumstances helps us to stay a step ahead of her many unpredictable consequences. That is why weather watching has become such an integral part of our daily activities.

How do we stay informed of late-breaking weather information? The most common method involves switching on an AM/FM radio or TV set, but we often receive more commercial announcements and entertainment than the immediately needed facts in the process. Do not be dismayed, however. Those general commercial broadcasters are only a tip of the weather information iceberg.

Additional sources available simply for the tuning include NOAA weather stations, amateur radio weather watch/emergency nets, and aircraft and marine weather services. These broadcasts and communications are on easily accessible frequencies in the shortwave HF and VHF ranges. The HF transmissions can be monitored on a shortwave radio, and VHF communications can be received on a scanner. Dedicated NOAA/weather band radios ranging from basic and inexpensive models to advanced tone alert models activated by signals preceding severe weather bulletins. They are readily available from electronics stores nationwide.

For individuals preferring more detailed and in-depth views, up-to-date weather facsimile (WEFAX) maps can be received directly from weather satellites transmitting in the 137-MHz range. WEFAX maps are also transmitted from coastal marine stations and can be received on several HF frequencies. Additional details on receiving setups are included later in this chapter and in Chapter 12 on home satellite receiving systems.

Whether traveling, vacationing, or protecting home base, rest assured several viable sources of late-breaking weather information are right at your fingertips. The key is being prepared and using the acquired knowledge in the most beneficial manner. House/AC powered receivers are fine during normal conditions, but battery powered units are more desirable for traveling and emergencies. Ideally, units capable of being powered from either source (and a dependable stock of fresh batteries) offer the greatest level of preparedness. Now let's take a closer look at some of the previously discussed weather information sources.

SEVEN NOAA WEATHER FREQUENCIES / CHANNELS	
(1) 162.550 MHz	(5) 162.450 MHz
(2) 162.400 MHz	(6) 162.500 MHz
(3) 162.475 MHz	(7) 162.525 MHz
(4) 162.425 MHz	

Figure 10-1 *The seven frequencies/channels used by NOAA for broadcasting weather information 24 hours a day. Radios and TV's capable of receiving these broadcasts often carry the NOAA logo.*

The NOAA Weather Radio System

Unquestionably, our country's most popular source of continuously available weather information is the NOAA weather radio system. This network of several hundred stations transmits frequently updated details on climates, temperatures, winds, and approaching storms, 24 hours a day on seven specially allocated channels in the 162-MHz range (Figure 10-1). The widespread popularity and vast coverage of these lifesaving broadcasts has resulted in the 162-MHz range becoming nicknamed the "NOAA Weather Band." Approximately 90 percent of our nation's area is presently within the 40 mile reception range of a NOAA weather transmitter site, and NOAA is striving to make that figure 100 percent. As a result, NOAA weather information is available to almost everyone at home, at work, or at play (Figure 10-2).

During severe weather, the National Weather Service, which was known as the U.S. Weather Bureau prior to 1970, supplies NOAA weather stations with special weather warning messages. The stations then transmit a 10-second, 1000-Hz tone to alert listeners prior to relaying messages. Less expensive NOAA weather receivers are not equipped to respond to the tone, but more expensive models will alert listeners by activating an internal miniature siren. Some models are even equipped with a flashing light to indicate an impending severe weather bulletin. NOAA's track record for saving lives by providing information precisely when needed is unsurpassed. Considering the low cost and numerous benefits of NOAA weather receivers, or scanners and receivers with NOAA weather coverage, we can comfortably state that every family or individual should have one for personal safety and security (see Figure 10-3). As further background on the NOAA weather system, the following information is presented from NOAA's public domain weather information booklet available from NOAA stations nationwide.

"NOAA Weather Radio is a service...of the National Oceanic and Atmospheric Administration (NOAA) of the U.S. Department of Commerce. As the "Voice of the National Weather Service," it provides continuous broadcasts of the latest weather information from local National Weather Service offices. Weather messages are repeated every 4 to 6 minutes and are routinely updated every 1 to 3 hours or more frequently in rapidly changing local weather or if a nearby hazardous environmental condition exists. Most stations operate 24 hours daily.

The regular broadcasts are specifically tailored to weather information needs of the people within the service area of the transmitter. For example, in addition to general weather information, stations in coastal areas provide information of interest to mariners and those in agricultural areas provide information of interest to farmers. Other specialized information, such as hydrological forecasts and climatological data may be broadcast.

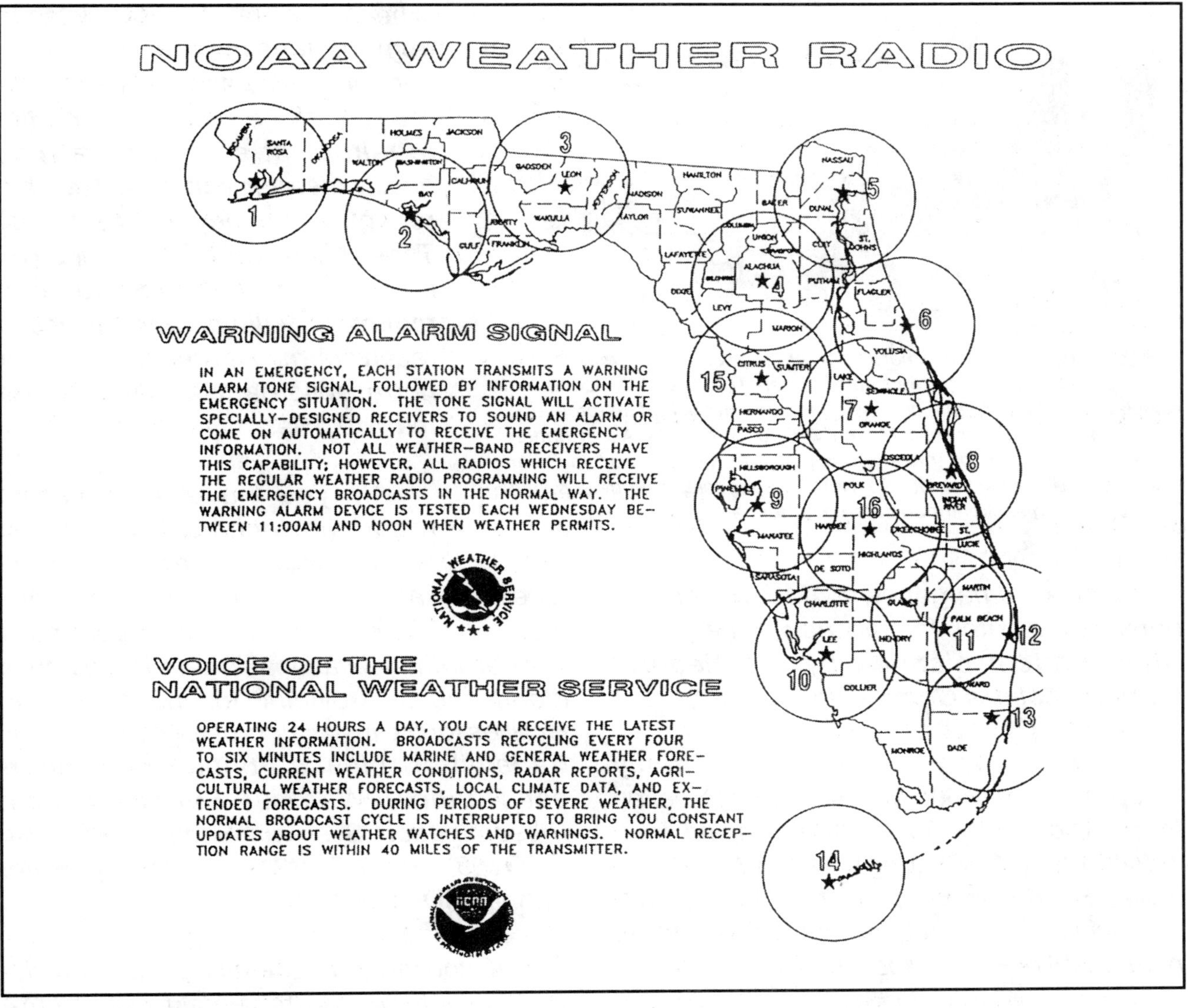

NOAA WEATHER RADIO
WARNING ALARM SIGNAL
IN AN EMERGENCY, EACH STATION TRANSMITS A WARNING ALARM TONE SIGNAL, FOLLOWED BY INFORMATION ON THE EMERGENCY SITUATION. THE TONE SIGNAL WILL ACTIVATE SPECIALLY-DESIGNED RECEIVERS TO SOUND AN ALARM OR COME ON AUTOMATICALLY TO RECEIVE THE EMERGENCY INFORMATION. NOT ALL WEATHER-BAND RECEIVERS HAVE THIS CAPABILITY; HOWEVER, ALL RADIOS WHICH RECEIVE THE REGULAR WEATHER RADIO PROGRAMMING WILL RECEIVE THE EMERGENCY BROADCASTS IN THE NORMAL WAY. THE WARNING ALARM DEVICE IS TESTED EACH WEDNESDAY BETWEEN 11:00AM AND NOON WHEN WEATHER PERMITS.
NATIONAL WEATHER SERVICE
VOICE OF THE NATIONAL WEATHER SERVICE
OPERATING 24 HOURS A DAY, YOU CAN RECEIVE THE LATEST WEATHER INFORMATION. BROADCASTS RECYCLING EVERY FOUR TO SIX MINUTES INCLUDE MARINE AND GENERAL WEATHER FORECASTS, CURRENT WEATHER CONDITIONS, RADAR REPORTS, AGRICULTURAL WEATHER FORECASTS, LOCAL CLIMATE DATA, AND EXTENDED FORECASTS. DURING PERIODS OF SEVERE WEATHER, THE NORMAL BROADCAST CYCLE IS INTERRUPTED TO BRING YOU CONSTANT UPDATES ABOUT WEATHER WATCHES AND WARNINGS. NORMAL RECEPTION RANGE IS WITHIN 40 MILES OF THE TRANSMITTER.

Figure 10-2 *State maps showing location and coverage area of NOAA weather transmitter sites are available from NOAA offices nationwide. Their telephone number is listed in your local phone directory.*

During severe weather, National Weather Service forecasters can interrupt the routine weather broadcasts and insert special warning messages concerning imminent threats to life and property. The forecaster can also add special signals to warnings that trigger "alerting" features in specially equipped receivers. In the simplest case, this signal activates audible or visual alarms, indicating that an emergency condition exists within the broadcast areas of the station being monitored and alerts the listener to turn up the volume and stay tuned for more information. More sophisticated receivers are automatically turned on and set to an audible volume when an alert is received.

In the most sophisticated alerting system, Weather Radio Specific Area Message Encoding (SAME), digital coding is employed to activate only those special receivers programmed for specific emergency conditions in a specific area, typically a county. SAME can activate specially equipped radio and cable television receivers and provide a short text message that identifies the location

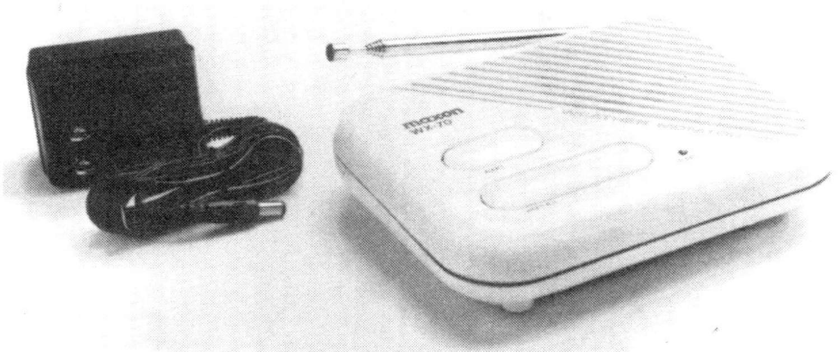

Figure 10-3 NOAA weather radios are available in several models and styles to fit every need and budget. They have the benefit of instant weather information, one button operation, and fail-proof use. Unit shown is the popular Maxon WX-70. It features AC or battery operation, choice of three warning signal modes, warning alert indicator, dial selection of all 7 NOAA frequencies/channels, pull-up antenna, and socket for an external antenna.

and type of emergency. SAME will be the primary activator for the new Emergency Alert System planned by the Federal Communications Commission.

NOAA Weather Radio currently broadcasts from 400 FM transmitters on seven frequencies in the VHF band, ranging from 162.400 to 162.550 Megahertz (MHz) in fifty states, Puerto Rico, the Virgin Islands, Guam, and Saipan. These frequencies are outside the normal AM or FM broadcast bands.

Special radios that receive only NOAA Weather Radio, both with and without special alerting features, are available from several manufacturers. In addition, other manufacturers are including NOAA Weather Radio as special features on an increasing variety of receivers. NOAA Weather Radio capability is currently available on some automobile, aircraft, marine, citizens band, and standard AM/FM radios as well as communications receivers, transceivers, scanners, and cable TV.

By nature and by design, NOAA Weather Radio coverage is limited to an area within 40 miles of the transmitter. The quality of what is heard is dictated by the distance from the transmitter, local terrain, and the quality and location of the receiver. In general, those on flat terrain or at sea, using a high quality receiver, can expect reliable reception far beyond 40 miles. Those living in cities surrounded by large buildings and those in mountain valleys with standard receivers may experience little or no reception at considerably less than 40 miles.

If you have a question regarding technical aspects of NOAA Weather Radio (such as reception and transmitter characteristics of a station) or are interested in becoming a partner with the National Weather Service in identifying or providing local funding and facilities for the installation of a Weather Radio transmitter, please contact your nearest National Weather Service Office, Dissemination Systems Section (Attn: W/OSO153), 1325 East-West Highway, Silver Spring, MD 20910.

If you have a question regarding the weather information broadcast over NOAA Weather Radio, please contact the local National Weather Service office that does the programming for the station or the National Weather Service, Warning and Forecast Branch (Attn: W/OM11), 1325 East-West Highway, Silver Spring, MD 20910."

VHF band marine handheld talkies have traditionally included weather channel reception. The recent concept of including similarly expanded weather frequency coverage in VHF band amateur radio equipment and talkies, newer CB radios, and portable AM/FM/TV receivers are proving to be a priceless asset. Keeping these facts in

Figure 10-4 *Small handheld CB units including NOAA weather band reception are a 2-for-1 bargain appealing to all survivalists. Unit shown here is the Maxon HCB-40WX, a deluxe 40 channel unit with push button selection of all 7 weather frequencies/channels.*

mind when shopping for various radios and communications equipment is definitely good thinking (see Figure 10-4).

Amateur Radio Weather Watches and Emergency Nets

Paralleling NOAA weather station reports and often exceeding their information with on-the-spot coverage during all types of abnormal weather are amateur radio's vast weather-watch groups and emergency networks. These groups/nets are typically recognized by names like SKYWARN (supplying data to NWS), AREC (Amateur Radio Emergency Corps), RACES (Radio Amateur Civil Emergency Service), and/or a related state's A, B or C emergency network. They are active on both VHF and HF bands/frequencies. The most widely known and easiest to receive local area networks are in the 145.00- to 147.30-MHz VHF range. HF nets usually handle and disseminate information on a larger (statewide or nationwide) scale; most are in the 3.800- to 4.000-MHz range. A secondary group of nets also operates in the 7.200- to 7.300-MHz range.

Precise frequencies used by nets on the previously mentioned 2-meter, 75-meter and 40-meter bands differ in various areas of the country; and they may also change frequencies in the future. We suggest checking the bulletin board at a local radio club or amateur radio dealer for presently active emergency net frequencies. If you are a licensed radio amateur, simply inquiring about an area's most active 2-meter FM repeater is all that is necessary. Alternately, merely scanning the previously mentioned 2-meter band segments during inclement weather usually proves ideal for spotting emergency nets.

Most radio amateurs are keenly interested in weather watching. Routinely, they switch their VHF FM transceiver or talkie to an area's repeater channel designated for emergency communications whenever unusual conditions occur. Some amateurs also set up ham radio communications equipment at local NOAA and NWS offices for collecting and disseminating information among weather spotters throughout a repeater's coverage area. As a result of combined efforts, the amateurs form a real time collection of eyes observing storms as they approach and grow or diminish in intensity. The amount of public service provided and knowledge shared via these nets is simply incredible. Indeed, true life stories of amateur radio's role in emergencies of all types are highlighted almost every month in the fraternity's leading magazines, CQ and QST. Both are sold at newsstands nationwide.

One example of amateur radio's role in weather watching recently occurred in your author's city when tornadoes moved across the state. When the National Weather Service first reported tornado formations approximately 100 miles west of the area, amateurs began gathering on the locally designated 146.880 MHz weather watch/ emergency repeater. Conversations regarding weather conditions in outlying areas and plans to activate SKYWARN facilities were soon replaced by emergency status activation of the repeater's network. As the storm front moved closer to the city, the pace of repeater activity increased. One amateur atop a distant mountain reported a wall cloud approaching from the southwest; another reported high winds and golf ball size hail at his location immediately ahead of the wall cloud. Meanwhile, amateurs on each side of the city gave reports that helped define the storm's overall size. When one of the tornadoes moved within visual range of the city, an amateur on a distant mountain described where the funnel cloud was located and its direction of travel. The report was confirmed by several more amateurs within the city. As the funnel cloud moved over the city and amateurs reported what was happening in their area, one could look toward that area and see the funnel cloud. Numerous reports of extreme winds, hail, and green sheet lightning (synonymous with tornadoes in the south) gave everyone monitoring an instantaneous mental picture of the tornado's progression. Meanwhile, stations behind the tornadic storm began reporting clearing weather. A second repeater was then specified for handling damage reports plus health and welfare information. During the whole scenario, reports were continuously being relayed to police, fire, weather, and other services.

Figure 10-5 Coverage of the full NOAA weather band is included in all popular models of amateur radio VHF transceivers and talkies. Any combination of frequencies/ channels can be stored in memories for instant recall.

The previous, and highly condensed story, is only one routine example of amateur radio's use and benefit during inclement weather and emergencies. It accurately illustrates the importance of getting an amateur radio license or at least monitoring amateur radio repeaters during any and all types of inclement weather. Remember, too, other weather information sources such as NOAA, FAA, and marine weather services can be monitored in the extended receiving range of an amateur radio VHF transceiver or talkie (Figure 10-5).

Aircraft and Marine Weather Stations

A creditable amount of weather information oriented toward aviation and marine use is transmitted on an around-the-clock basis, and it is also available to anyone monitoring related frequencies. As you learned with NOAA and amateur radio activities, VHF transmissions usually relate to local areas and HF transmissions usually relate to weather on a larger geographic scale.

First, and most well known, are weather conditions announced at nearby airports. This information is transmitted in the AM aircraft band between 118 and 136 MHz, with most airports using 119 to 122.500 MHz for weather data. The easiest route for determining a specific frequency used in a particular area is by simply scanning the above mentioned ranges, then loading the active channel(s) into your receiver's memories. Try it! Scan and memorize techniques work great and monitoring aircraft weather is quite interesting. Generally speaking, aircraft weather stations can be received between 10 to 20 miles from an airport, or farther if you are atop a high mountain.

The first time you scan the "air band" you will probably find a local airport's basic information frequency. Listen closely and note the auxiliary frequencies they specify for weather, arrivals, and departures. You may even note mention of clearance delivery frequencies or FAA and military frequencies in the 325- to 375-MHz range. Assuming your receiver tunes those often overlooked frequencies, you have a direct line to FAA weather data. Once again, the key to spotting active frequencies is scanning. In this case, using the 325- to 375-MHz AM mode rather than FM mode for reception. Some popular frequencies used throughout the country are 122.00 MHz (aviation weather) and 342.50 plus 344.60 MHz (FAA).

Weather information for aircraft on long flights or covering large geographic areas can also be received on HF bands. Most notable are New York radio on 3.485 MHz, 10.051 MHz and 13.270 MHz (listen on the hour and half hour for East Coast weather) and Oakland Radio on 6.680 and 8.828 MHz (listen five minutes past the hour and half hour for West Coast weather).

Although a dedicated VHF weather service for mariners is not available at the present time, the Coast Guard broadcasts weather data several times daily on marine radio channel 22A (157.100 MHz). The Coast Guard usually announces the upcoming weather broadcasts on marine channel 16 (165.800 MHz), the distress and calling channel used by boaters. The Coast Guard's weather broadcasts are in the FM mode and are most informative for people living near the shore. On a larger geographic scale, listen for Coast Guard weather on the following HF frequencies: 3.023, 5.320, 6.506, 11.195 and 15.081 MHz.

High power coastal stations providing land-based communications with ships at sea also transmit weather information and forecasts on the HF bands. Some of the transmissions are by voice using SSB mode; some are weather facsimile pictures transmitted by data modes. A listing of the most well-known weather announcements by SSB follows.

KMI in California transmits on 4.402 and 13.083 MHz at 0000 GMT and 1200 GMT.

WOM in Florida transmits on 4.363, 13.092 and 17.242 MHz at 1300 GMT and 2300 GMT.

WOO in New Jersey transmits on 4.387 and 8.750 MHz at 1200 GMT and 2200 GMT.

Additional weather broadcasts on various schedules include the following coastal stations: WLO in Alabama on 2.572 MHz, WAK in Louisiana on 2.482 and 4.419 MHz, KQP in Texas on 2.530 MHz, and KOW in Washington state on 2.522 MHz.

Weather facsimile maps are transmitted by KMI at 20 minutes past odd hours on 8.087 MHz and by WOO at 20 minutes past even hours on 8.051 MHz. Additional WEFAX stations and transmissions will be discussed in the following part of this chapter.

Whether inland or shore based, aircraft and maritime weather services are available to everyone. They fill a noticeable gap in available information and are always worthy of consideration. Remember, too, HF air and marine stations can be received anywhere you travel. Simply tune to lower frequencies (below 10 MHz) during evening hours and upper frequencies (above 10 MHz) during daytime hours for optimum "skip" and strong signal reception.

Weather Fax: Weather Pictures and Maps

An impressive selection of weather fax (WEFAX) pictures are transmitted daily on HF, VHF, and satellite TV frequencies. They are quite an informative medium for serious weather watchers. These weather maps are similar to those you see on evening television newscasts. Indeed, they are often exact "originals" used in many weathercasts. Equipment used to receive weather pictures ranges from economical interface units that connect between a receiver and home computer to complex stand-alone systems with more features than an individual would ever use. All varieties reproduce pictures with good resolution.

Weather fax pictures are transmitted and received on a line-by-line basis. First, a transmitting station scans the photo to be transmitted, converting it into tiny picture elements (pixels) making up each line. Each pixel is represented by an electrical voltage. These voltages are then converted to tone variations and transmitted. At the receiving station, a fax demodulator converts the signal from tones to voltages which are applied to a stylus or print head which makes light and dark marks on a piece of paper. Each line is reconstructed pixel by pixel, and the full picture is reassembled/printed line by line. Most pictures are transmitted at the rate of 60, 90, 120, or 240 lines per minute. Weather maps are views of the earth from many miles up and typically show cloud cover, temperature, barometric pressure, and wind direction. Many maps include overlays of country boundaries and states. By comparing maps received over a period of several hours, one can actually see the movement of storms and fronts. Many weather maps are incredibly accurate and also capable of a resolution to less than one mile. With practice, one can even recognize clouds moving over their home. On clear days, spotting area lakes and streams is common.

As previously mentioned, the three main sources of weather fax maps are HF transmissions from land-based marine and meteorological stations, dedicated VHF/UHF-band weather satellites, and WEFAX retransmission via commercial 3.7-4.2 GHz satellites. Some of the most impressive HF WEFAX transmissions emanate from coastal stations like NAM in Virginia and WLO in Alabama. NAM transmits on 3.357, 8.080 and

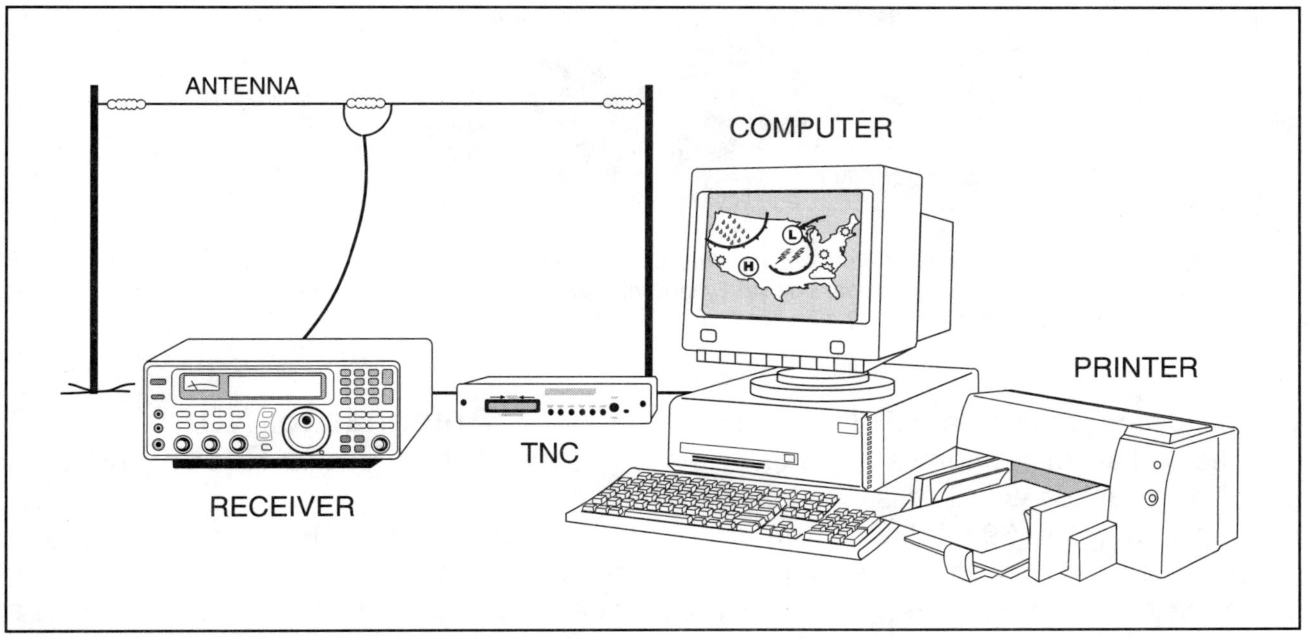

Figure 10-6 *Outline of equipment used to receive WEFAX transmissions. NOAA's National Environmental Satellite, Data and Information Service (NESDIS) overlays maps on cloud photos faxed to Earth from weather satellites. The results are transmitted by WEFAX on high frequency (HF) bands. WEFAX signals transmitted between 2 and 20 MHz are easy to monitor and display. You'll need a stable shortwave receiver with an appropriate antenna, a terminal node controller (TNC) and a computer or fax device to display or print weather maps.*

10.865 MHz using SSB mode 24 hours daily. They also transmit on 20.015 MHz from 1200 to 2100 GMT and on 16.410 MHz from 0900 to 2100 GMT daily. WLO transmits on a varying schedule on 6.850 and 17.447 MHz, SSB mode, daily. The WEFAX format of each is 120 lines per minute.

The equipment required for HF WEFAX reception is illustrated in Figure 10-6. The shortwave receiver must be capable of SSB mode reception and exhibit good frequency stability for consistently solid picture copy. It should also be connected to an outdoor wire antenna between 20 and 50 feet in length to ensure medium to strong signal reception. A WEFAX converter or computer interface unit (called a TNC) connects to the receiver's earphone/external speaker socket. The converter or computer, with WEFAX software installed, then connects to a monitor for real time displays and/or to a printer for hard copies. Dedicated converters such as the Universal Model M-8000 shown in Figure 10-7 are ideal, because the operator need not be a computer "whiz" to use it. Simply connect a cable to the receiver's audio output, another cable to a video monitor and a printer for hard copies, dial the fax format used, and tune the receiver to a fax frequency.

The most well known and often used weather satellites are operated by the National Oceanic and Atmospheric Administration, and transmit in the 137.5- to 137.65-MHz range using wideband FM mode. Most weather watchers tune in NOAA 9 on 137.620M Hz, NOAA 10 on 137.500 MHz, NOAA 11 on 137.620 MHz, and/or NOAA 12 on 137.500 MHz. The same converter and display units employed for HF are utilized; however, a high gain helix antenna and

Figure 10-7 *The Universal M-8000 is a stand-alone unit for copying weather fax pictures received via a shortwave HF or VHF radio or home satellite TV system (TVRO). Unit connects between the radio's audio output and a video monitor and/or a printer. Details in text.*

wideband receiver with GaAsFET preamplifier is used in lieu of a shortwave receiver. A second group of weather satellites also transmit in the 1691-MHz range, and antennas plus downconverters that permit 137-MHz receivers to tune 1691 MHz are employed.

Finally, a good variety of weather maps are relayed by commercial satellites carrying network TV programs. These pictures may be received by owners of home satellite receiving systems operating in the 3.7- to 4.2-GHz range. In this case, a shortwave receiver is typically connected to the satellite receiver's "wideband" output for tuning hidden frequencies between transponders. Additional information is presented in this book's chapter on home satellite/TVRO systems.

Weather plays a significant role in all of our lives and is a paramount consideration for survival in all areas. Armed with the knowledge presented in this chapter, one should be well equipped to face any circumstances experienced in the present or future.

Antennas for NOAA Weather Radio Receivers

NOAA Weather Radio receivers, both crystal-controlled and tunable, capable of picking up the National Weather Service's continuous weather transmissions, are available at a variety of prices. The price is usually proportional to the sensitivity and quality of the receiver. Consult your nearest Weather Service office for information on the exact broadcast frequency.

The range and clarity of reception is dependent upon the height of the transmitting antenna, the terrain effects, and the distance from the transmitting site.

To enhance the quality of reception at greater distances, an antenna attached to the receiver is recommended. As a rule, the use of a sensitive quality receiver and antenna assures more reliable and clearer reception.

NOAA Weather Radio transmits a vertically polarized signal; therefore, a vertically polarized receiving antenna that will receive signals in the 130 - 175 MHz range is recommended.

If the receiving antenna can be positioned at a minimum height of 100 feet at locations near the fringe of optimum reception (approximately 40 miles), a ¼ wave length ground plane antenna is recommended. Prices range between $6 and $20.

For distances beyond the fringe reception area, a ½ wave length dipole antenna with a gain of not less than 4 to 6 dB should be used. This type is also available as a directional antenna which amplifies reception manifold. These range in price from $50 to $100 or more.

There are a number of ground plane antennas ($6 to $20 range) suitable for use on cars, boats, and indoors.

The diagrams of antenna systems are designed for receivers equipped with external antenna jacks or fittings. Smaller receivers

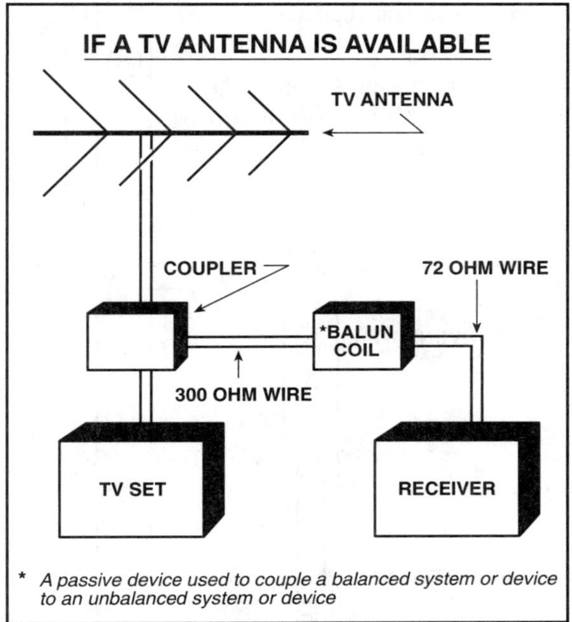

* *A passive device used to couple a balanced system or device to an unbalanced system or device*

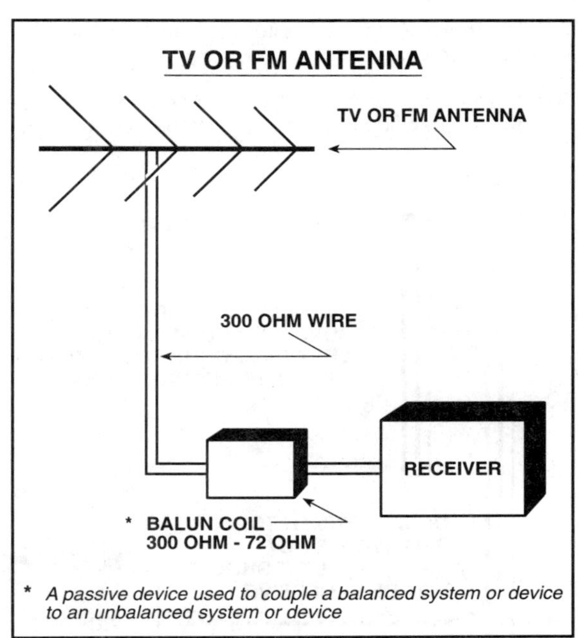

* *A passive device used to couple a balanced system or device to an unbalanced system or device*

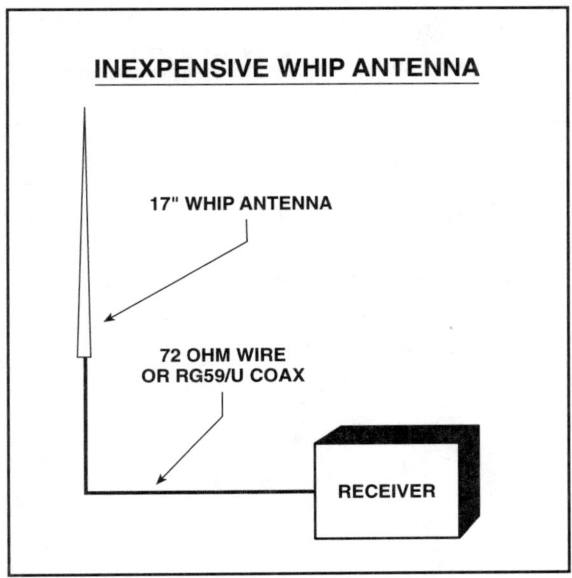

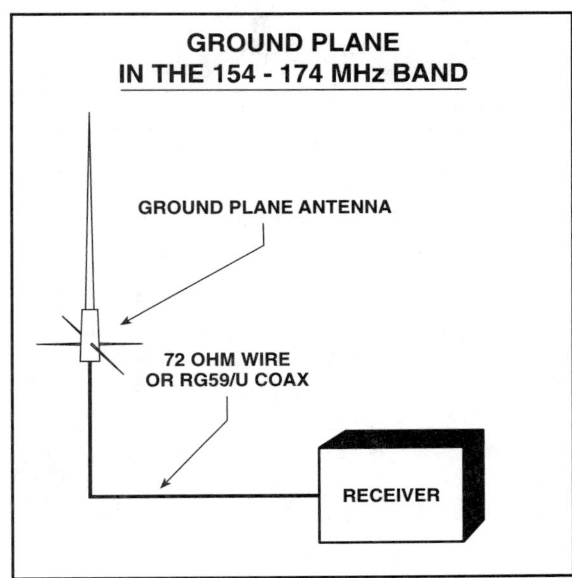

using a telescoping whip antenna must be wired inside (to antenna and to ground). It is recommended that this modification be done by a professional or a knowledgeable amateur with experience in radio electronics.

It is recommended that you consult your local electronic equipment supplier regarding these specifications prior to purchasing a specific antenna for any receiver.

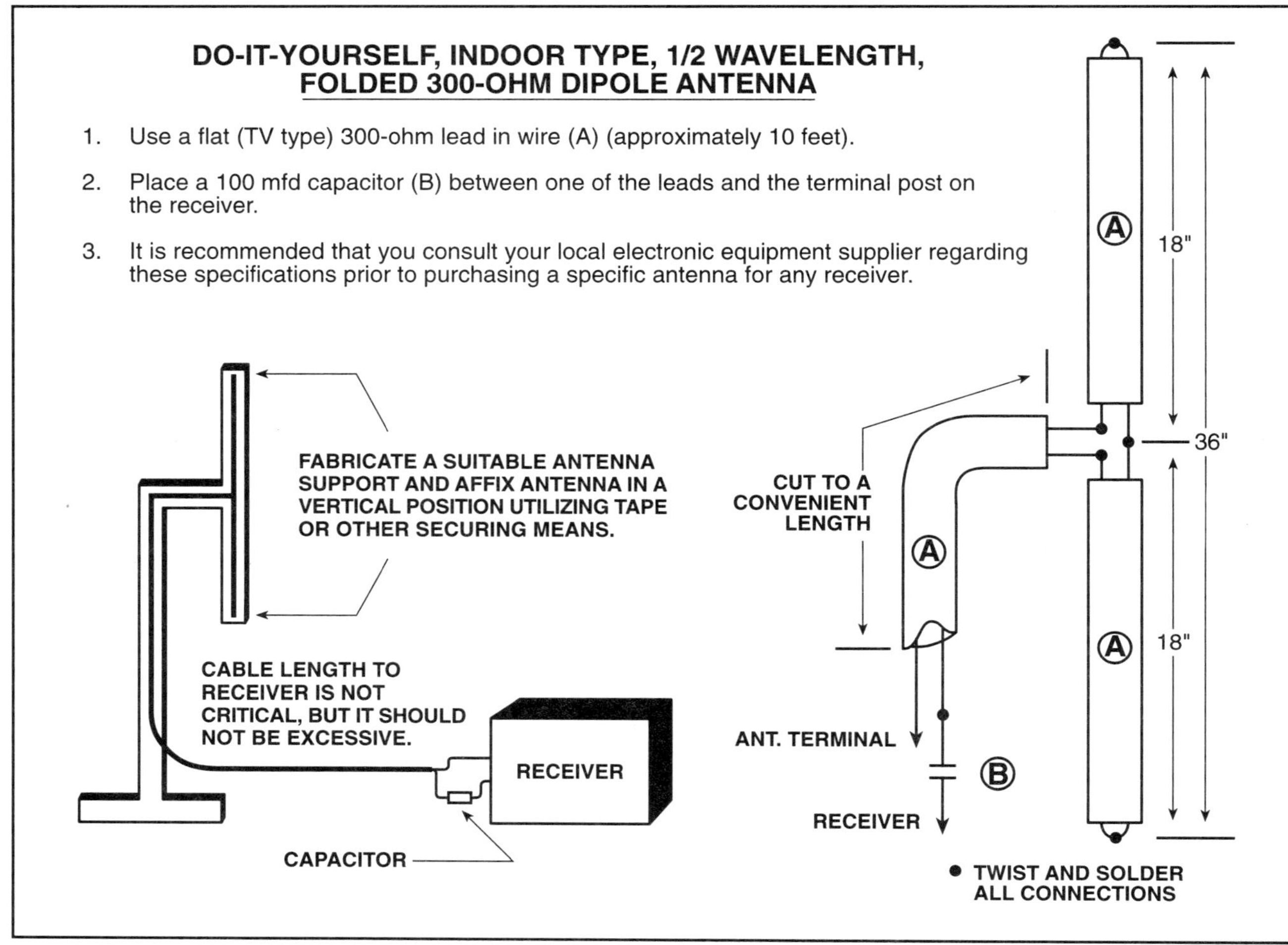

DO-IT-YOURSELF, INDOOR TYPE, 1/2 WAVELENGTH, FOLDED 300-OHM DIPOLE ANTENNA
1. Use a flat (TV type) 300-ohm lead in wire (A) (approximately 10 feet).
2. Place a 100 mfd capacitor (B) between one of the leads and the terminal post on the receiver.
3. It is recommended that you consult your local electronic equipment supplier regarding these specifications prior to purchasing a specific antenna for any receiver.
FABRICATE A SUITABLE ANTENNA SUPPORT AND AFFIX ANTENNA IN A VERTICAL POSITION UTILIZING TAPE OR OTHER SECURING MEANS.
CABLE LENGTH TO RECEIVER IS NOT CRITICAL, BUT IT SHOULD NOT BE EXCESSIVE.
RECEIVER
CAPACITOR
CUT TO A CONVENIENT LENGTH
A
ANT. TERMINAL
B
RECEIVER
A
18"
36"
A
18"
• TWIST AND SOLDER ALL CONNECTIONS

Antennas for VHF Receivers

The performance of radio receivers to tune to the National Weather Service's continuous weather transmissions may be enhanced considerably with the addition of a fixed antenna suitably exposed. The range of reception is affected by such factors as the height of the transmitting and receiving antenna, intervening terrain, and the sensitivity of the receiver. For maximum performance, the use of a sensitive (better than 10 micro-volts) receiver and a fixed antenna is suggested.

National Weather Service VHF-FM transmissions are vertically polarized. A vertically polarized receiving antenna that will receive signals in the 130-175 MHz range is recommended.

If the receiving antenna can be positioned at a minimum height of 100 feet at locations near the fringe of optimum reception (approximately 40 miles), a ¼ wave length ground plane antenna is recommended. Prices range between $6 and $20.

For distances beyond the fringe reception area, a ½ wave length dipole antenna with a gain of not less than 4 to 6 dB should be used. This type is also available as a directional antenna which improves reception considerably. These range in price from $50 to $100 or more.

There are a number of ground plane antennas ($6 to $20 range) suitable for use on cars, boats and indoors.

Use of low-loss coaxial cable to connect the antenna to the receiver is recommended.

An indoor, ½ wave length folded dipole antenna can be made as shown in the diagram.

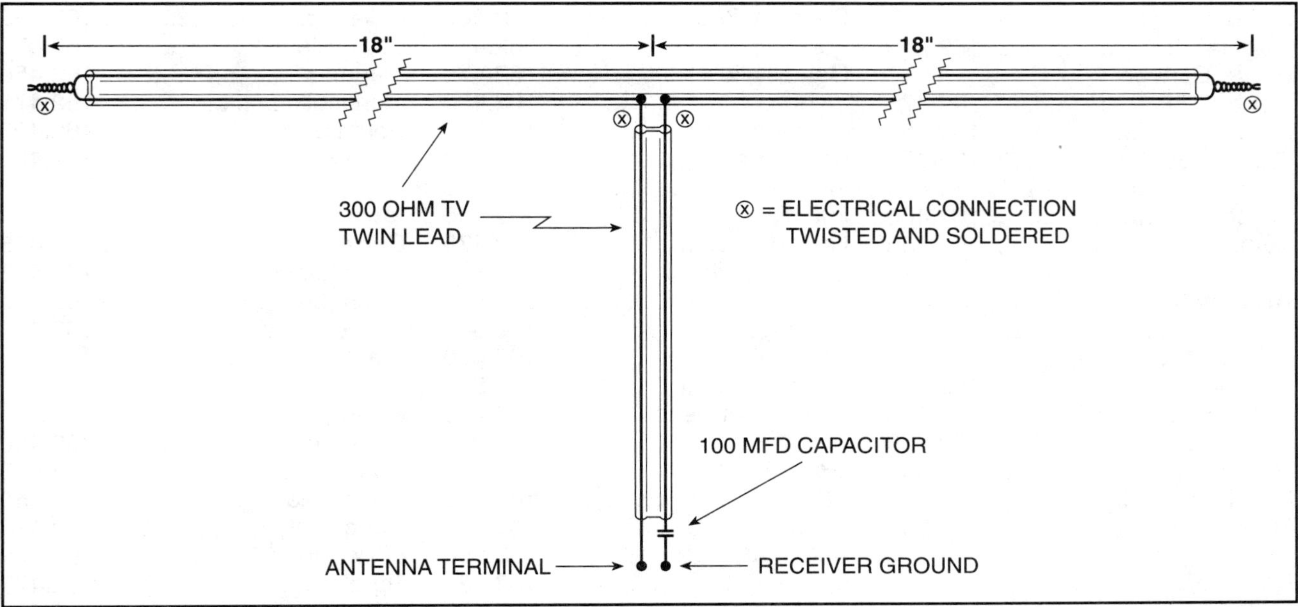

NOAA Weather Transmitter Sites

Alabama
Anniston 162.475
Birmingham 162.550
Demopolis/Linden 162.475
Dozier 162.550
Florence 162.475
Fort Payne 162.500
Huntsville 162.400
Louisville 162.475
Mobile 162.550
Montgomery 162.400
Tuscaloosa 162.400

Alaska
Anchorage 162.550
Cordova 162.550
Craig 162.400
Fairbanks 162.550
Haines 162.400
Homer 162.400
Juneau 162.550
Ketchikan 162.550
Kodiak 162.550
Nome 162.550
Seward 162.550
Sitka 162.550
Valdez 162.550
Wrangell 162.400
Yakutat 162.400

Arizona
Flagstaff 162.400
Grand Canyon (Hopi Point) 162.475
Phoenix 162.550
Porter Mountain 162.400
Show Low (Porter Mt) 162.400
Tucson 162.400
Yuma 162.550

Arkansas
Fayetteville 162.475
Fort Smith 162.550
Gurdon 162.475
Jonesboro 162.550
Little Rock 162.550
Mountain View 162.400
Star City 162.400
Texarkana 162.550

California
Bakersfield 162.550
Coachella 162.400
Eureka 162.400
Fresno 162.400
Grass Valley (Wolf Mt.) 162.400
Lindsay 162.550
Los Angeles 162.550
Monterey 162.550
Pt Arena/Ukiah 162.550
Redding 162.550
Sacramento 162.400
San Diego 162.400
San Francisco 162.400
San Luis Obispo 162.550
Santa Barbara 162.400

Colorado
Alamosa 162.475
Colorado Springs 162.475
Denver 162.550
Fort Collins 162.450
Grand Junction 162.550
Greeley 162.400
Mead/Longmount 162.475
Pueblo 162.400
Sterling 162.400

Connecticut
Hartford 162.475
Meriden 162.400
New London 162.550

Delaware
Lewes 162.550
Salisbury, MD 162.475

Florida
Belle Glade 162.400
Daytona Beach 162.400
Fort Myers 162.475
Gainesville 162.475
Inverness 162.400
Jacksonville 162.550
Key West 162.400
Melbourne 162.550
Miami 162.550
Orlando 162.475
Panama City 162.550
Pensacola 162.400
Sebring 162.500
Tallahassee 162.400
Tampa 162.550
West Palm Beach. 162.475

Georgia
Athens 162.400
Atlanta 162.550
Augusta 162.550
Baxley 162.525
Chatsworth 162.400
Columbus 162.400
Macon 162.475
Pelham 162.550
Savannah 162.400
Valdosta 162.500
Waycross 162.475
Waynesboro 162.425

Guam
Agana 162.400

Hawaii
Hawaii (Kulani Cone) 162.550
Kauai (Kokee) 162.400
Maui (Mt. Haleakala) 162.400
Oahu (Mt. Kaala) 162.550
Oahu (Kai) 162.400

Idaho
Boise 162.550
Brundage Mt. 162.475
Lewiston 162.550
Pocatello 162.550
Twin Falls 162.400

Illinois
Champaign 162.550
Chicago 162.550
Marion 162.425
Peoria 162.475
Rock Island/Moline 162.550
Rockford 162.475
Springfield 162.400

Indiana
Bloomington 162.400
Evansville 162.550
Fort Wayne 162.550
indianapolis 162.550
Louisville, KY 162.475
Marion 162.450
Monticello 162.475
South Bend 162.400
Terre Haute 162.400

Iowa
Cedar Rapids 162.475
Des Moines 162.550
Dubuque 162.400
Sioux City 162.475
Waterloo 162.550

Kansas
Chanute 162.400
Colby/Goodland 162.475
Concordia 162.550
Dodge City 162.475
Ellsworth 162.400
Topeka 162.475
Wichita 162.550

Kentucky
Ashland 162.550
Bowling Green 162.400
Covington 162.550

Elizabethtown 162.550
Hazard 162.475
Lexington 162.400
Mayfield 162.475
Paintsville 162.525
Pikeville 162.400
Somerset 162.550

Louisiana
Alexandria 162.475
Baton Rouge 162.400
Buras 162.400
Lafayette 162.550
Lake Charles 162.400
Monroe 162.550
Morgan City 162.475
New Orleans 162.550
Shreveport 162.400

Maine
Caribou 162.525
Dresden 162.475
Ellsworth 162.400
Falmouth 162.550

Mariana Islands
Saipan Mra 162.550

Maryland
Baltimore 162.400
Hagerstown 162.475

Massachusetts
Boston 162.475
Hyannis (Camp Edwards) 162.550
Mt. Greylock 162.525
Worcester 162.550

Michigan
Alpena 162.550
Detroit 162.550
Flint 162.475
Grand Rapids 162.550
Hesperia 162.475
Houghton 162.400
Marquette 162.550
Onondaga 162.400
Oshtemo 162.475
Sault Ste Marie 162.550
Traverse City 162.400

Minnesota
Bemidji 162.425
Detroit Lakes 162.400
Duluth 162.550
International Falls 162.550
Mankato 162.400
Minneapolis/St. Paul 162.550
Rochester 162.475
Roosevelt 162.475
St. Cloud 162.475
Thief River Falls 162.550
Willmar 162.475

Mississippi
Ackerman 162.475
Booneville 162.550
Bude 162.550
Columbia 162.400
Gulfport 162.400
Hattiesburg 162.475
Inverness 162.550
Jackson 162.400
Meridian 162.550
Oxford 162.400

Missouri
Camdenton 162.550
Columbia 162.400
Hannibal 162.475
Hermitage 162.450
Joplin 162.550
Kansas City 162.550
Sikeston 162.400
Springfield 162.400
St. Joseph 162.400
St. Louis 162.550

Montana
Billings 162.550
Butte 162.550
Glasgow 162.400
Great Falls 162.550
Harve (Squaw Butte) 162.400
Helena 162.400
Kalispell 162.550
Miles City 162.400
Missoula 162.400
Plentywood 162.475

Nebraska
Bassett 162.475
Grand Island 162.400
Holdrege 162.475
Lincoln 162.475
Merriman 162.400
Norfolk 162.550
North Platte 162.550
Omaha 162.400
Scottsbluff 162.550

Nevada
Elko 162.550
Ely (Cave Mtn.) 162.400
Las Vegas
(Boulder City) 162.550
Reno 162.550
Winnemucca 162.400

New Hampshire
Concord 162.400

New Jersey
Atlantic City 162.400

New Mexico
Albuquerque 162.400
Carlsbad 162.475
Clovis 162.475
Des Moines 162.550
Farmington 162.475
Hobbs 162.400
Las Cruces 162.400
Ruidoso 162.550
Santa Fe 162.550

New York
Albany 162 550
Binghamton 162.475
Buffalo 162.550
Elmira 162.400
Kingston 162.475
New York City 162.550
Riverhead 162.475
Rochester 162.400
Stamford 162.400
Syracuse 162.550
Watertown 162.475

North Carolina
Asheville 162.400
Cape Hatteras 162.475
Charlotte 162.475
Fayetteville 162.475
New Bern 162.400
Raleigh/Durham 162.550
Rocky Mount 162.475
Wilmington 162.550
Winston Salem 162.400

North Dakota
Bismarck 162.475
Dickinson 162.400
Fargo 162.475
Jamestown 162.550
Minot 162.400
Petersburg 162.400
Williston 162.550

Ohio
Akron 162.400
Bridgeport 162.525
Caldwell 162.475
Cleveland 162.550
Columbus 162.550
Dayton 162.475
Lima 162.400
Sandusky 162.400
Toledo 162.550

Oklahoma
Clinton 162.475
Enid 162.475
Lawton 162.550
McAlester 162.475
Oklahoma City 162.400
Ponca City 162.450
Tulsa 162.550

Oregon
Astoria 162.400
Bend/Redman 162.475
Brookings 162.550
Coos Bay 162.400
Eugene 162.400
Klamath Falls 162.550
Medford 162.400
Newport 162.550
Pendleton 162.400
Portland/Estacada 162.550
Roseburg 162.550
Salem 162.475

Pennsylvania
Allentown 162.400
Clearfield 162.550
Erie 162.400
Harrisburg 162.550
Johnstown 162.400
Philadelphia 162.475
Pittsburgh 162.550
State College 162.475
Towanda 162.550
Wellsboro 162.475
Wilkes-Barre 162.550
Williamsport 162.400

Puerto Rico
Maricao 162.550
San Juan 162.400

Rhode Island
Providence 162.400

South Carolina
Beaufort 162.475
Charleston 162.550
Columbia 162.400
Conway/Myrtle Beach 162.400
Cross 162.475
Florence 162.550
Greenville 162.550
Sumter 162.475

South Dakota
Aberdeen 162.475
Huron 162.550
Pierre 162.400
Rapid City 162.550
Sioux Falls 162.400

Tennessee
Bristol 162.550
Chattanooga 162.550
Cookeville 162.400
Jackson 162.550
Knoxville 162.475
Memphis 162.475
Nashville 162.550
Shelbyville 162.475
Waverly 162.400

Texas
Abilene 162.400
Amarillo 162.550
Austin 162.400
Beaumont 162.475
Big Spring 162.475
Brownsville 162.550
Byran 162.550
Corpus Christi 162.550
Dallas 162.400
Del Rio 162.400
El Paso 162.475
Fort Worth 162.550
Galveston 162.550
Houston 162.400
Laredo 162.400
Lubbock 162.400
Lufkin 162.550
Odessa/Midland 162.400
Paris 162.550
Pharr 162.400
San Angelo 162.550
San Antonio 162.550
Sherman 162.475
Tyler 162.475
Victoria 162.400
Waco 162.475
Wichita Falls 162.475

Utah
Logan 162.400
Milford/Cedar City 162.400
Navajo Mountain 162.550
Salt Lake City 162.550
St. George (Utah Hill) 162.475
Tooele (South Mt) 162.475
Tooele (Vernon Hills) 162.400
Vernal 162.400

Vermont
Burlington 167.400
Marlboro 162.425
Windsor 162.475

Virginia
Heathsville 162.400
Lynchburg 162.550
Norfolk 162.550
Richmond 162.475
Roanoke 162.475

Washington, DC
(Manassas) 162.550

Virgin Islands
St. Thomas 162.475

Washington
Neah Bay 162.550
Okanogan (Tunk Mt.) 162.525
Olympia 162.475
Seattle 162.550
Spokane 162.400
Wenatchee 162.475
Yakima 162.550

West Virginia
Beckley 162.550
Charleston 162.400
Clarksburg 162.550
Gilbert 162.475
Hinton 162.425
Moorefield 162.400
Spencer 162.500
Sutton 162.450

Wisconsin
Adams 162.400
Green Bay 162.550
La Crosse 162.550
Madison 162.550
Menomonie 162.400
Milwaukee 162.400
Park Falls 162.500
Sister Bay 162.425
Wausau 162.475

Wyoming
Casper Mountain 162.550
Cheyenne 162.475
Lander 162.475
Sheridan 162.500

CHAPTER ELEVEN

Emergency and Alternate Power Sources

In the same way food and water are necessary for "human fuel," electrical energy is required for powering various types of radios and communications equipment. Indeed, even the most expensive scanner or the most elaborate two-way system is of little benefit without a reliable source of power for use during both normal times and emergencies. As we are also aware, the availability of commercial power is unpredictable at best during any significant emergency. Fortunately, most emergencies and commercial power outages are less than 72 hours in length, and batteries are a logical source of energy during such times. More viable alternatives are necessary, however, for longer term outages and for people living in remote areas. The emphasis then shifts to alternate energy sources for stand-alone operation for true self-reliance. This chapter reviews a variety of alternate energy sources suitable for powering both communications equipment and home utilities for an independent lifestyle. A number of concepts and ideas will be discussed, as various areas of the country differ in their available resources, and each person's needs are unique.

Although seldom realized, alternate energy sources have been with us for many years. Batteries, for example, have been used for powering flashlights and portable radios since the early 1900's. Basic non-rechargeable batteries are suitable for occasional applications, but their life is limited. Rechargeable batteries overcome that dilemma, but a source of energy for recharging them must be available. Maintaining a stock of freshly charged batteries satisfies short-term needs, but alternatives are necessary to answer long-term needs.

A logical first consideration many people select involves using a gasoline or diesel powered generator to "back up" commercial power facilities. This concept has merit, provided an adequate supply of fuel is maintained on hand for the engine. The key is keeping your own storage tanks of fuel prepared and minimizing generator engine noise. Extreme measures? Not at all; radio and TV stations maintain large diesel powered generators and underground fuel tanks for emergency and long-term use. An increasing number of families in rural and backwoods areas are taking note and following suit. Being self-reliant in this unpredictable world is truly a step worthy of everyone's consideration. Gasoline or diesel powered generators are not the ultimate answer for all applications, however, and several more sources of alternate energy are available at a reasonable cost. These alternatives include solar panels and wind or water driven generators. They will be discussed later in this chapter.

Light Duty Cells and Batteries

Our discussion on batteries begins with a brief review of basic 1.5-volt cells used in flashlights and portable radios. We use the term "basic" to describe non-rechargeable items, incidentally, and "cells" to specify single low voltage components. A battery is comprised of a bank or "battery" of single cells, and usually produces 9 to 12, or more, volts output. Batteries and cells are alike in the respect that they store energy in chemical form and convert it to electrical form when used. Cells (and batteries of cells) can be separated into two categories or types: non-rechargeable and rechargeable. Both types use chemicals for their electrolyte and dissimilar metals (such as copper and zinc) for their positive and negative electrodes.

Cells used to power flashlights and portable electronic equipment are categorized as types AA, AAA, C and D. Each delivers 1.5-volts DC, assuming they are "fresh" or fully charged. The types differ only in physical size and amount of current they deliver. Larger cells obviously deliver more current than smaller cells (see Figure 11-1). Variations in the battery's chemical composition also influences the amount of current each size cell can produce. As an example, alkaline batteries are well known for "longer life" (i.e., greater current capacity) than generic dry cells. Now, let's add some figures to the previous facts and visualize how they relate to powering electronic equipment.

Figure 11-1 *Standard 1.5-volt cells are available in AAA, AA, C, and D sizes. Difference in size determines milliamp-hour ratings, or how long they will power a particular device.*

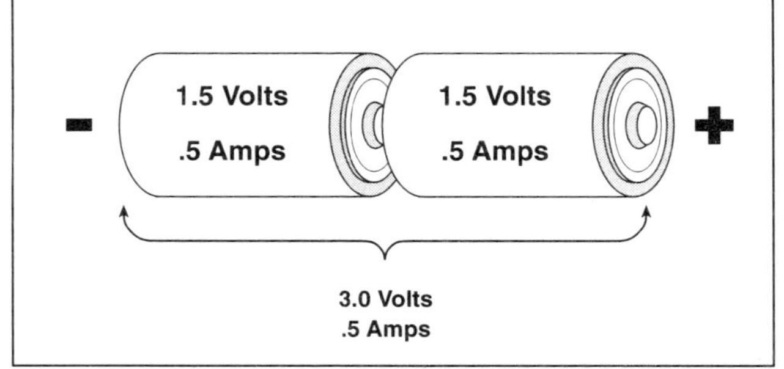

Figure 11-2 *Arrangement for connecting cells in series so their voltage adds while the current is the same as each individual cell. Note there is only one unbroken path for current flow.*

Alkaline AA cells are typically rated to deliver between one-half ampere (500 milliamperes) and six-tenths of an amp (600 mA) at 1.5 volts. AAA alkaline batteries typically deliver 250 mA (one quarter amp), C cells deliver 1.2 amps (1200 mA), and D cells deliver 1.5 amps (1500 mA). The common 9-volt battery used in many portable radios, incidentally, is rated to deliver 90 mA (.09 amp). As a comparison, generic carbon cells typically deliver one-third (.33%) to one-quarter (.25%) less current than alkaline cells of the same size; accordingly they are less expensive and exhibit a shorter lifespan. In order to understand the previous facts and apply them to real-world situations, we should briefly discuss some basic electronic theory applicable to all types of cells, batteries, and equipment.

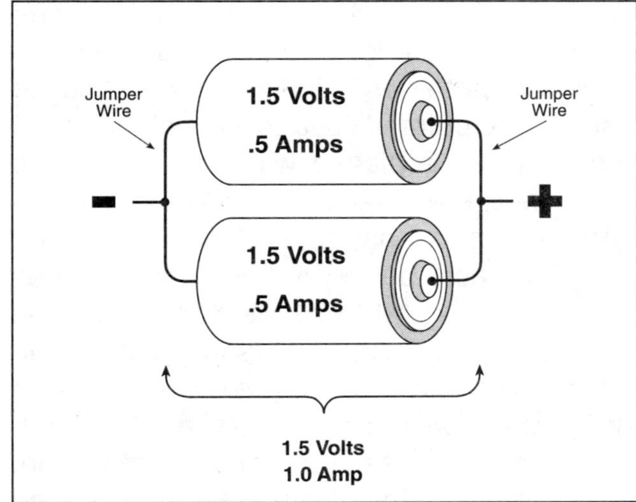

Figure 11-3 *Arrangement for connecting cells in parallel so their currents add while voltage is same as each individual cell. Note there is more than one path for current flow. Break one path, and energy is still available (which is not the case in a series arrangement).*

A logical first consideration with batteries is how long they will power a particular unit, so let's begin by visualizing "hourly factors." Remember the previously mentioned 500 mA rating for AA cells? That is their "full charge" (as in "full" on an automobile's fuel gauge), and the cell's total one hour rating. If the cells power a device requiring 1.5 volts at 500 mA, it will become fully discharged ("empty") in one hour. If it powers a device requiring 250 mA, the cell will last two hours. If the cell powers a device requiring 50 mA (typical for a pocket tape recorder), approximately 10 hours of operation can be expected. If a D cell could be substituted (probably by using clip lead jumpers), approximately 30 hours operation could be expected (1500 mA ÷ 50 mA per hour used = 30 hours).

On the other hand, recorders, handheld talkies, etc. require more than one cell, and that brings our next point to focus. When cells are connected/wired in series (Figure 11-2), their voltage adds and their current rating (mA per hour) stays the same as each individual cell. For example: 8-AA cells = 8 x 1.5 (volts per cell), or 12 volts at 500 mA. When cells are connected/wired in parallel (Figure 11-3), their voltage stays the same as that of each cell and their currents add. For example: 3-AA cells = 1.5 volts at 3 x 0.5 (amps per cell), or 1.5 amperes. Connecting cells or batteries in parallel thus provides more hours of operation for a particular device. Finally, cells may be connected in a combination series-parallel arrangement for obtaining both greater voltage and current (See Figure 11-4).

How can we determine how much current (per hour) a device requires for operation? If you have a volt-ohm-milliamp meter, simply connect its leads in series with one cell's terminal and measure the current drawn when the device is operating. That is its milliamp-hour rating. Alternatively, voltage and current ratings of electronic devices and equipment are usually stamped/marked on their cases or listed in their instruction

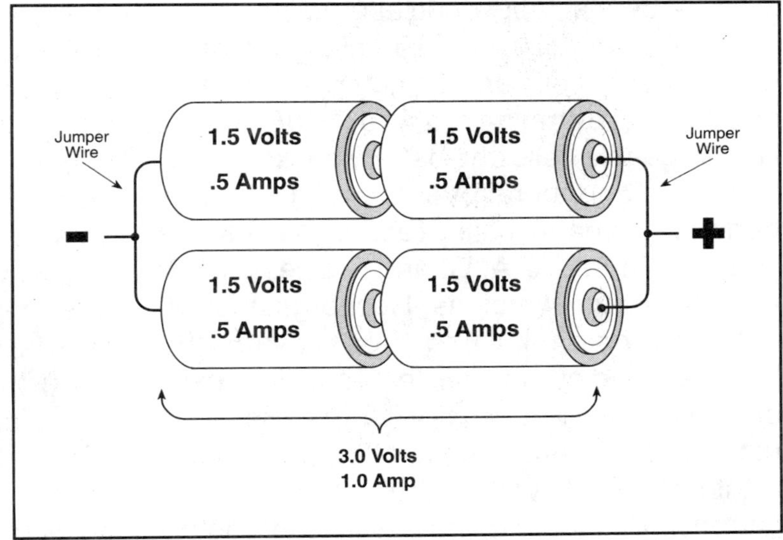

Figure 11-4 *Combination of series and parallel connected cells. Voltage is total of each series "string." Current is total of all parallel-connected "strings."*

manual's specifications. With practice, you will be able to accurately predict how long various devices/units will operate on various size cells and batteries. Really! Try it!

The previous facts also apply to AC power and equipment/appliances; however, some products are marked only with a voltage and wattage rating. The formula here is voltage times amperes equal watts, or conversely, watts divided by volts equals amperes (i.e., current). As an example, a small VCR is typically marked "24 watts." When connected to 120 volts, that equates to 0.2 amperes (200 mA). Again, practice use of "electrical math" makes on-the-spot understanding easy. Simply remember there is no "hidden magic" in electronics; everything equates right down to the milliampere.

Rechargeable Cells and Batteries

Generic dry cells and batteries are economically priced but limited in the respect they have but one life to give to related equipment; consequently, "rechargeable" cells become an appealing alternative. When properly cared for, nickel-cadmium and other rechargeable cells and battery packs can be used and recharged up to 1000 times. Rechargeable cells and battery packs pay for their cost, and more, over the long run. Like generic "primary" cells, rechargeables are available in AA, AAA, C, and D sizes plus 9-volt and 12-volt versions. Rechargeable 12-volt battery packs like those used with handheld radios and talkies usually consist of 10 series-wired cells. Why 10? Because rechargeables are usually rated at 1.2 volts rather than 1.5 volts at full charge. Their current/milliampere rating is akin to, or slightly less than, alkaline batteries. Rechargeables differ, however, in their discharge rate. They deliver full output until almost discharged, then drop off or "die" quickly. Herein lies the secret for proper care of rechargeable nickel-cadmium batteries. Ideally, they should be used to the point of voltage drop off and then recharged or replaced with a backup pack until/while the original pack is recharged. Nickel-cadmium batteries, incidentally, differ from lead-acid batteries in the respect they can be left or stored in full discharge condition without adverse effects. In fact, rechargeable batteries sold with talkies are usually shipped uncharged. Two more points regarding nickel-cadmium batteries are worthy of mention. First, avoid using them after "discharge begins" (when output begins dropping off), as it will cause reverse current flow and eventually short-circuit cells. Second, avoid frequent "top off" charging of partially discharged cells/packs, as it promotes a "memory effect." This effect is realized or characterized by cells/packs becoming accustomed to receiving a nice recharging at "half full" and consequently "dying" at that point rather than holding a full charge. The solution? Use two or three packs and interchange them as needed. Occasionally "topping off" a cell is okay; just do not allow it to become a habit.

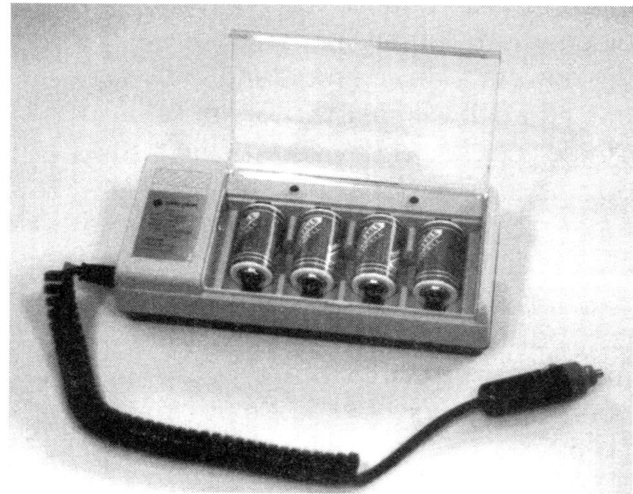

Figure 11-5 *A novel Ni-Cad battery charger for recharging "D," "C," "AA" and 9-volt batteries from an auto cigarette lighter socket or any 12-volt DC source. Available from many electronics stores or preparedness mail order houses.*

At this point, I should mention the Ray-O-Vac Company recently introduced an impressive line of AAA, AA, C and D size reusable alkaline cells that look to be very attractive for powering portable communications equipment. The cells exhibit a long shelf life and more current than an alkaline or a nickel-cadmium equivalent cell. Their cost is comparable to an alkaline cell and they can be recharged 25 to 100 times. Unlike nickel-cadmium cells, Ray-O-Vac's Renewable alkaline batteries can be "topped off" as much as desired without "memory" problems. In fact, topping off their charge actually extends their life. Many people use communications talkies, portable shortwave radios, etc. on an occasional basis and want "fresh batteries" for immediate use. In such cases, Ray-O-Vac Renewables are an ideal answer. A photo of the Ray-O-Vac cells and their matching charger is shown in Figure 11-6.

Another rechargeable energy source is gel cells and batteries, and they are not prone to memory effects as are rechargeable nickel-cadmium cells. Indeed, gel cells may be an ideal form of hassle-free batteries. They can be topped off and readied for emergencies as needed, plus deep cycled (within reason) without adverse effects. Gel cells are slightly

Figure 11-6 *Ray-O-Vac line of renewable alkaline cells looks quite promising. These cells exhibit long shelf life, deliver high current, and can be recharged 25-100 times.*

Figure 11-7 *Gel-cell batteries are available in various physical sizes, voltages, and current ratings. They have proven to be worry-free energy sources for powering small communications equipment.*

heavier than nickel-cadmium cells, and they are usually capable of delivering high current (Figure 11-7).

Gel cells and nickel-cadmium cells are recharged at one-tenth their milliampere hour rating for eleven hours (the last hour is used to "put back slightly more than we took out" and compensate for heat losses). Alternately, they may be "quick charged" at their full milliampere hour rating for one hour and five minutes. Usually, slightly more voltage than a cell or battery delivers at full charge is required to provide the current needed for charging.

Bringing all the previous facts together, we can logically say rechargeable cells and batteries are ideal for day-to-day preparedness and use during anticipated short-term emergencies. Alkalines exhibit longer life, but conventional types cannot be recharged. Ideally, having a good stock on hand is the best bet.

The area of light duty cells and batteries is continuously expanding, and new designs are appearing on the market each year.

Armed with the previously discussed facts, however, you should be able to intelligently evaluate new types of cells and batteries and determine how they might serve your own needs.

Heavy Duty Batteries

We often need a method of storing a large quantity of energy for on-demand use, and high current rechargeable batteries are necessary to fill these requirements. The most commonly known examples here are 6-volt golf cart storage batteries. These heavy-duty commercial batteries are capable of retaining 350 to 500 amperes, and are known as lead-acid batteries because their plates are lead compounds and their electrolyte is sulfuric acid. Lead acid batteries are available in two basic types: completely sealed and "refillable." The latter type is recognized by removable caps for adding distilled water to each cell.

Figure 11-8 Two battery fuel gauges that measure the amount of energy in the battery by measuring the battery's top voltage which in turn depicts the total percentage of charge. These are used with sealed lead-acid batteries and regular deep-cycle lead-acid types.

Unlike nickel-cadmium batteries, lead-acid batteries can be continuously "topped off" so their full current is always available for use. Overcharging is prevented by a voltage regulator system.

Storage batteries are not known for holding their full charge over a long period of time like nickel-cadmium and gel cells, but that is not their intended purpose. Factually, a storage battery is best visualized as a form of "electrical storage tank." We might even compare it to an automobile's fuel tank that stores gasoline, and it too is refillable. In other words, we can use part of the battery's current for powering electrical devices. The current drawn out is then replaced from a DC source. The battery should not be allowed to run down below 50% of its rated capacity to prevent damage to the unit.

Automobile storage batteries are not the proper type of lead-acid battery for permanent battery banks. Your auto battery can be used in the event of an emergency by running jumpers from your auto's battery to communications equipment; however, a few pertinent facts should be considered. First, avoid draining the battery to the point that not enough current remains to start the vehicle. Second, low voltage and high current cannot be transferred more than five or ten feet without a significant voltage drop. The battery and the communications equipment need to be near each other and connected by heavy (large) wires. Finally, remember battery acid is explosive around flames and spills burn clothing or skin. Additionally, high currents heat wires and can literally melt soft metallic objects such as rings and expansion watch bands. Always exercise care and caution when working with high current batteries.

Deep cycle lead-acid batteries used for trolling motors and interior/cabin lighting in RVs or boats are perfect for powering electronic equipment, provided they are mated with a good charging system. Unlike auto batteries, deep cycle types have higher current ratings and are especially made for repeated deep discharge cycles. They are not designed, however, to deliver intermittent bursts of high current. Some popular applications of deep cycle batteries include mating them with generators and solar cells for charging, and using the full system to power remotely located radio equipment. These systems can also be "sized up" and used with inverters to power mountain cabins, or "sized down" to produce portable energy units.

Proper maintenance of deep cycle batteries involves ensuring they are clean, charged, and the water/acid level is above interior plate tops. By nature, these heavy-duty batteries are often exposed to extreme cold (which cracks cases and drains electrolytes), extreme heat (which boils electrolytes and causes cell short circuits), and dirt (which creates high resistance paths so the battery cannot deliver full current). Keep the batteries' casings and tops clean, scrape the terminals and connectors shiny clean at least annually, keep the electrolyte level above the batteries' plates, keep the batteries well charged, and they will reward you with dependable service in emergencies.

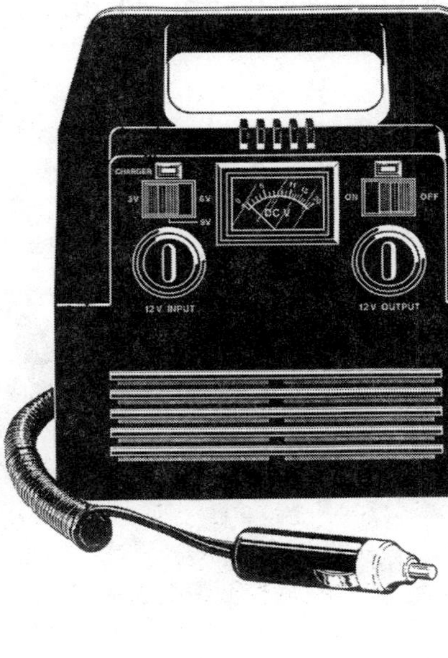

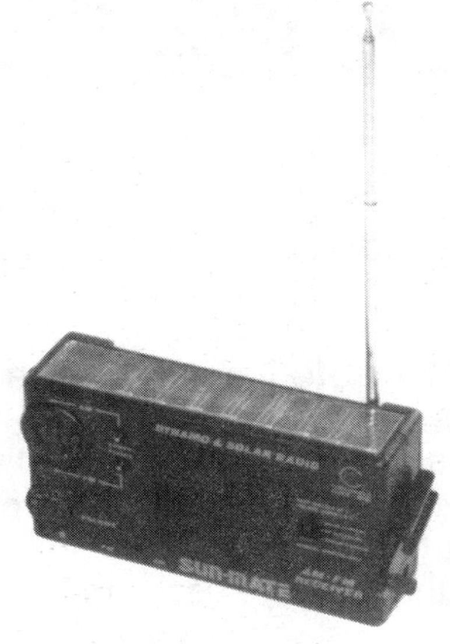

Figure 11-10 *The 4-way powered radio is great for emergencies or everyday use. It can operate from any one of four power sources. Good for camping and outdoor use and should be included in your survival gear.*

Recently, several types of portable energy systems built around sealed lead-acid batteries have become popular for on-the-spot use. These units are compact and versatile, and usually consist of a 6 to 16 ampere, 12-volt DC battery plus a built-in regulator and chargers for home and auto use. The incorporation of a sealed lead-acid battery into an emergency power system is attractive because it is capable of delivering enough surge current to jump start an auto or power smaller communications equipment for several days. One good example of these "refillable energy tanks" is shown in Figure 11-9.

Figure 11-9 *Portable energy system produced by Valor Enterprises, Inc. Unit sports a sealed lead-acid 12 volt/ 7 ampere battery plus energy monitoring circuitry. Item is capable of powering communications equipment for many hours and even jump-starting an auto.*

Figure 11-11 Powerport 149, Portable Power Module, furnishes 155 volts AC and 12 volts DC in a carry-around bag.

Portable Energy Items

Many preparedness dealers and catalogs carry some inexpensive worthwhile items that make a lot of sense. Several products that we have used are illustrated in this section. The first is an unusual small radio, AM/FM, that is powered from four sources.

This product can be powered by a set of solar cells built into the top of the case. The radio also has a built-in dynamo that will charge the internal battery and will also use common dry cells that can be used with a 110-volt AC wall transformer. This 4-way small radio sells for under $30.00 and works very well from any of the power sources (see Figure 11-10)

Hand-Portable Power Module

The Powerport 149 is a compact 4" x 4.5" x 6" unit, weight 9 pounds. This unit provides 140 watts of 115 volts AC (surges to 200 watt) and up to 20 amps of 12-volt DC power. For example, a 2-way handheld radio will run up to 80 hours. A padded bag with accessory pouches protects the unit from impact and makes transport a breeze. At the heart of this rechargeable power supply is a sturdy 12 volt, 9 amp hour sealed gel-cell battery. The Powerport 149 can be charged in your vehicle through the cigarette light plug without the engine running. It is also equipped with a fully automatic 115-VAC wall charger which can be left plugged in without fear of overcharging the battery. The Powerport 149 will run and charge many devices, such as handheld radios, test equipment, emergency lighting, handheld GPS receivers, laptop computers and small electric hand tools (see Figure 11-11).

Solar Powered Ni-Cad Battery Charger

An all-in-one solar charger that will recharge AAA, AA, C and D cell Ni-Cad

Figure 11-12 A perfect little Ni-Cad solar battery charger. Sells for under $25.00. Charges all sizes of Ni-Cad batteries.

Figure 11-13 *Professional grade power inverter made by Trace Engineering Company and available through Alternative Energy Systems in Redway, CA. Unit converts 12 or 24 volts DC/battery energy into 120 or 240 volt AC power. Note socket on left panel for DC input and wall-type 120-volt AC socket on right panel. This particular inverter handles loads up to 1500 watts.*

batteries and also extend the life of most alkaline batteries. One solar charger does it all! Built-in blocking diode prevents discharge of batteries when there is no sun. This charger will charge most new flat prismatic-type batteries and includes a voltage test and sun meter that indicates battery voltage and when the batteries are being charged. The charger measures 7" x 3" x 2 ½", is placed in a sunny window for full charging, and will pay for itself many times. Perfect for emergency power or everyday use (see Figure 11-12).

Converters and Inverters

At this point, let's pause briefly to explain two energy-related items associated with battery/DC power systems: converters and inverters. Simply explained, a converter is used to change one level of DC or AC voltage to another level of DC voltage. An inverter changes one level of DC voltage to another level of AC voltage. In other words, a converter lets us use home AC energy to power battery-operated equipment, and an inverter lets us use batteries to power AC-operated equipment.

One easily recognized example of a converter is a small DC power supply for a CB radio. The unit plugs into a regular 120-volt AC outlet, and its internal electronics transform and rectify the output to 12 volts DC for power to the radio. Transformation steps the voltage down by a particular ratio. In the case of the CB power supply, the ratio is 10:1. Rectification changes AC (which alternates between positive and negative polarity many times each second) to DC (which has specific positive and negative terminals). Two other forms of converters are "wall adapters," miniature DC power supplies used to power portable radios on 120 volts; and multi-voltage DC "cigarette lighter" adapters for powering 3- or 6-volt DC recorders, etc. in automobiles.

Inverters are not "common items" like converters. However, an easily understood example is a device used on boats and crafts for powering 120-volt AC televisions, shavers, food processors, lights, etc. from 12 volts DC. Since direct current cannot be directly applied to a step-up transformer, it must first be electronically "chopped up" or changed to

Figure 11-14 *A small commercial power supply consisting of a 300-watt inverter with a 12 volt / 7 ampere-hour gell-cel sealed battery. This unit will power 110 VAC radios and other devices in the 300-watt range.*

Figure 11-15 This Power Pony™ mini-generator measures only 10 by 10 by 11 inches (H,W,D), weighs only 13 pounds, and goes almost anywhere. It delivers 110 volts AC and 12 volts DC for operating low power communications equipment, a computer, etc.

quasi-AC. The inverter changes DC to AC and then performs a 1:10 step-up of voltage. Inverters are usually more complex in design and more expensive than converters. In the past, inverters were relatively inefficient and only capable of handling light loads (100 to 200 watts). Newer versions available today are incredibly efficient (85 percent of their power input in watts is available as output in watts, with only 15 percent dissipated as heat). They are also made in models ranging from 500 watts to 8,000 watts. The latter are capable of providing AC power for a house or small office from a large bank of storage batteries. Most of today's communications equipment is capable of direct 12-volt DC operation (or easily modified to accept 12-volt DC power). If a unit is not 12-volt DC ready, however, using an inverter is a logical alternative. A professional grade inverter is shown in Figure 11-13.

Motor Driven Generators

In the absence of commercial electrical power, gasoline or diesel engine driven generators are popular for producing large amounts of energy. Generators are available in numerous styles and sizes, ranging from 400 watts to over 25,000 watts, and they are ideal for powering large communications systems or even complete dwellings. Engine driven generators are an attractive source of alternate electrical power, as they do not rely on wind, water movement, or sunlight for operation. However, their liquid fuel, noise, and maintenance, require in-depth study and planning for long-term benefits. Let's consider some of these points.

First, gasoline engine generators must run at high RPM's to deliver full-rated output (typically 3,600 RPM). As such, they consume a large amount of fuel per hour and are

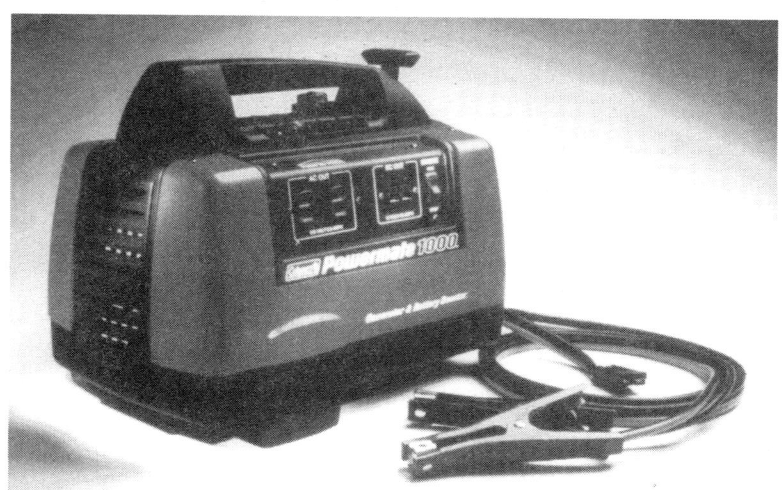

Figure 11-16 The Coleman Powermate 1000™ is a high power, 850 watt portable generator for more demanding energy requirements. Unit delivers 120 volts AC or 12 volts DC, measures 12 by 14 by 18 inches, and weighs 26 pounds.

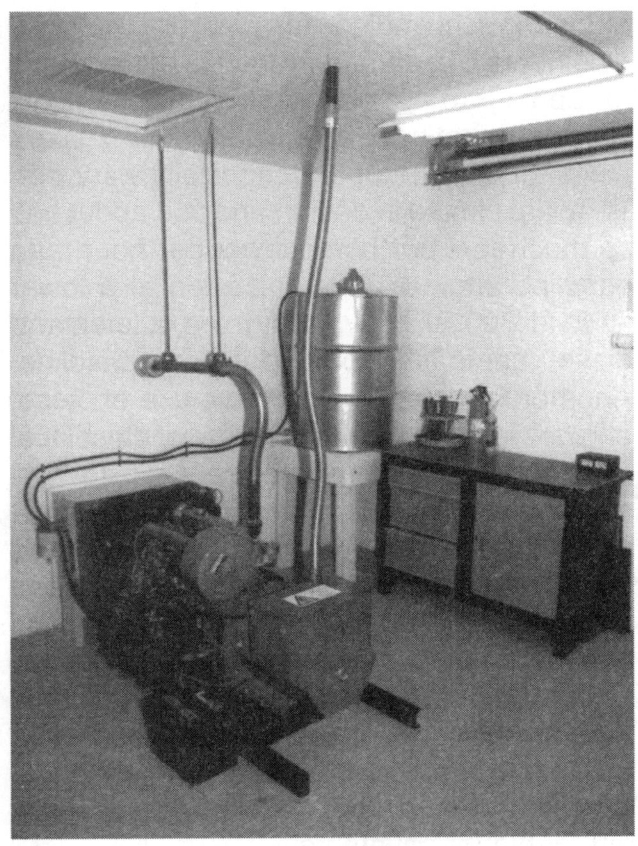

Figure 11-17 Addressing high energy demands for extended power outages is this generator from Imperial Diesel. It consumes half the fuel of a gasoline engine, requires less maintenance, and powers both communications equipment plus household needs.

really not suitable for continuous use. Gasoline engines operating near full throttle also require a goodly amount of maintenance and servicing. In other words, they are good for intermittent use or for light duty backup during brief power outages but are not cost-effective for daily use. One route around that drawback is pairing a generator with a heavy-duty inverter and bank of storage batteries. That way, the generator's energy can be stored and used as needed rather than being operated sporadically and having half of its energy wasted. Factually, generators operated to near their maximum wattage rating (average, not peak) exhibit longer life than sporadically used generators. If all of a generator's energy is not needed at a particular time, diverting it to an electrical storage system (batteries) is a logical choice.

Noise and fuel consumption may seem like minor points, and they may be minor in temporary situations, but they can become significant over a 24-hour, or longer, period. Noise alerts others to your resources, possibly inviting vandalism. Dual mufflers, sound-proofed sheds or basements, etc., may be used to minimize noise and yield peace of mind. Burying a 55-gallon drum, or larger, for storing your own supply of fuel is also worthwhile. Keep in mind the larger the engine and generator, the more fuel it consumes. If your budget is limited and your emergency power needs light (e.g., operating 100 to 500 watts of communications gear or powering the blower of a gas heater for a few hours), a gasoline engine driven generator may serve nicely. Alternately, a diesel engine driven generator is worthy for long-run use. Gasoline engines are available in various sizes and price ranges. Intelligent shopping will result in selecting one best suited to your needs. A diesel driven generator will have a serviceable life four times that of a gasoline driven unit.

Figure 11-18 This Imperial Diesel generator is designed to power entire dwellings plus communications equipment and more on a long-term basis.

Figure 11-19 *An example of a diesel generator shed designed to provide maximum protection for the generator and its fuel in addition to ensuring environmental safety and quiet operation. The large battery bank is housed for proper temperature environment.*

One of the most versatile and unique mini-generators we have seen is the table top Power Pony™ shown in Figure 11-15. This 350-watt unit is available from The House of Generators, and the cost is quite low. The Power Pony™ produces 110 volts AC at up to 3.2 amps and 12 volts DC at 4 amps, making it ideal for powering small communications gear, scanners, a TV, desk lamp, etc. on a moment's notice. The generator has an electronic ignition, recoil starter, and runs eight hours on a gallon of gasoline when delivering 300-watts output.

A larger and more powerful generator in the portable category is the Coleman Powermate 1000™ shown in Figure 11-16. This unit produces 850 watts (120 volts AC at up to 7.1 amps or 12 volts DC at up to 70 amps) for powering large communications systems or some home utilities such as a furnace blower or refrigerator. The unit has a recoil starter, an Auto Throttle™ that sets the RPM to match the load, measures 12 by 14 by 18 inches H,W,D, and weighs 26 pounds.

Diesel engine driven generators are well known by professionals for their rugged, maintenance-free operation and long life. Indeed, diesel engines are capable of 40,000 plus hours of operation which is comparable to running an auto one to two million miles. Diesels do not have an ignition system, spark plugs, or carburetor to periodically wear out. They are simple in design and use about half as much fuel per horsepower/per hour than gasoline engines. They also run at a lower RPM (1,800 RPM, typically), are quieter, and diesel fuel is less expensive than gasoline. Another important factor in the use of diesel engines is the longer storage life of diesel fuel over gasoline. With proper care, diesel fuel can be stored in underground tanks for at least five years and can be revived after extended storage by adding a few gallons of kerosene to a large tank of diesel fuel. For long runs and high power use, diesel engine driven generators are the optimum choice. Until recently, the cost of diesel power has been quite high. Times are changing, however, and imported diesel engines are now surprisingly affordable for private use.

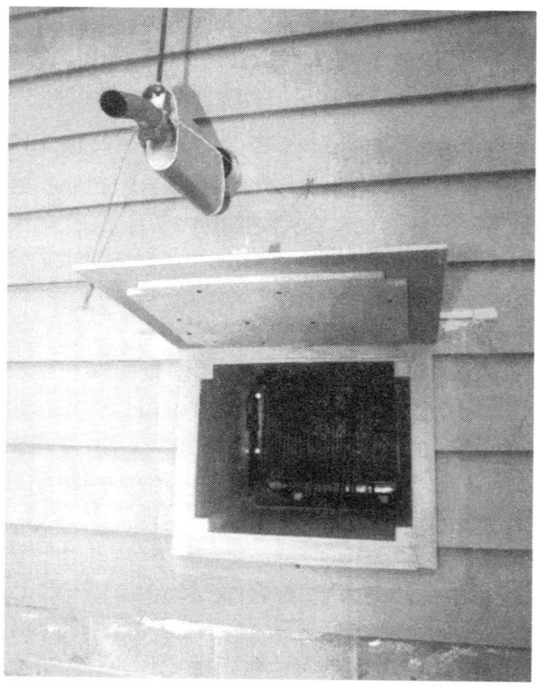

Figure 11-20 *The diesel generator shed's exhaust and radiator openings are designed for optimum efficiency and quiet operation.*

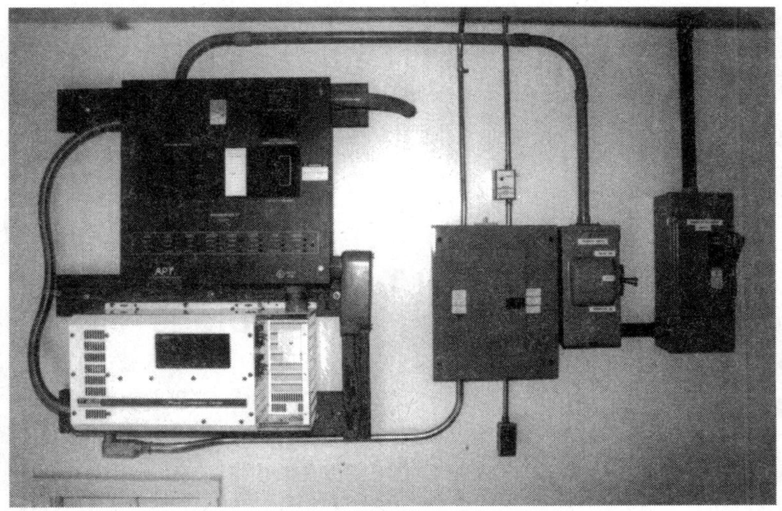

Figure 11-21 *Photo showing three control panels for switching from solar power to generator power. From right to left: First is a 2-position switch, 1) generator on, 2) generator off line. Second is a 3-position selective switch, 1) solar power, 2) generator, 3) commercial power grid. Third, a control panel with circuit breakers for generator building, fan, lights, heater, 4) main solar power control center, and 5) 4-kW, 24-volt DC inverter to provide 110-volt AC power.*

Some of the more impressive units we have seen are being marketed by Imperial Diesel. Several models ranging from 8 kW to 23 kW are available (some models are shown in Figures 11-17 and 11-18). These German-designed diesels are supplied with British made Stamford Newage brushless alternators; both are rugged and reliable systems.

The choice of a smaller intermittent duty generator or a larger continuous duty generator, using a gasoline or diesel engine will ultimately be determined by your needs and budget. General knowledge of types, however, ensures a good buying decision.

Fuel for Alternate Power Sources

An often overlooked element in the selection of alternate power sources using gasoline or diesel generators is the proper selection and long-term storage of a fuel source. Some simply think several five gallon cans of gas stored in a safe place is all that is needed—nothing could be further from the truth!

Gasoline stored for over three to five months will disable any generator in short order. Today's unleaded fuel will form gummy residues and varnish-like films that will clog up passages in fuel lines, carburetors, and injectors in generators.

Any fuel, be it gasoline or diesel, is made up of many organic compounds. These organic compounds are constantly changing over time, becoming new compounds that change the characteristics of fuel. The same molecules that make up the best parts of gasoline can react with oxygen and other elements in the environment to form new molecules that build up to form residues and films that will cause problems in the generator engine.

Some fuels are treated with oxidation inhibitors to allow them to be stored for up to

Figure 11-22 *The Monitor 650 unit plugs into any wall receptacle to accurately show AC voltage and frequency (cycles) of the power source. This unit is also used to check the operation of the auxiliary generator and possible inverter output.*

three months without generating excessive deposits.

Other fuels have no inhibitors at all. In any case, storage of fuel requires some special consideration in order for the fuel to be usable in time of need.

One way to avoid most of the problems with bad fuel is to completely drain the fuel tank and fuel lines; however, this procedure poses a number of problems. First, it is virtually impossible to get every drop of fuel out of the system by simply draining the fuel. Second, draining the fuel exposes bare metal in the tank and fuel system to air and moisture, which together can form rust and corrosion and can allow gaskets to dry out, crack, and shrink leading to fuel leaks when the system is refilled. For these reasons, draining the fuel is not the solution.

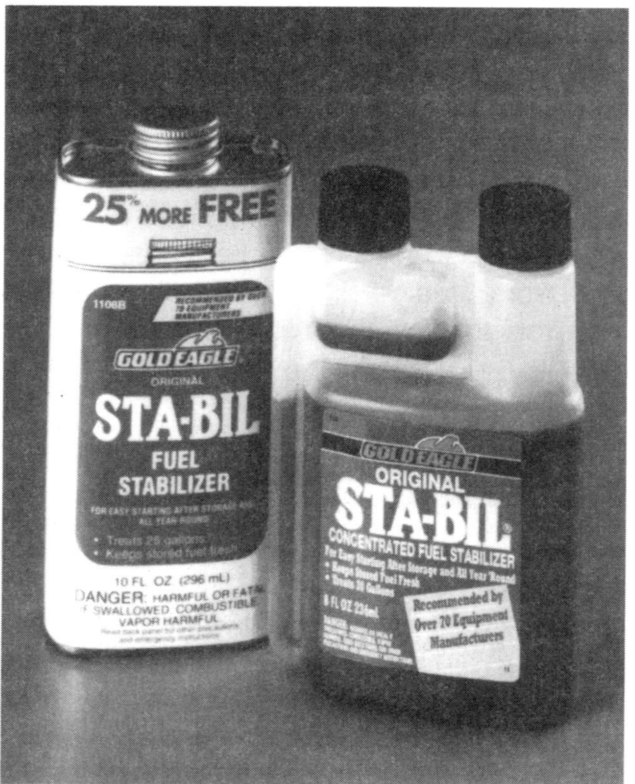

Figure 11-23 STA-BIL®, a widely distributed fuel additive to increase the fuel storage life of gasoline. A diesel fuel additive is also available from STA-BIL®.

Fuel stored in cans and tanks will also oxidize with many of the "bad-acting" unstable molecules remaining in the solution, In other words, stored fuel needs treatment also.

Fuel Treatments for Long-Term Storage

We are fortunate that several companies market quality fuel treatments. One product we have used for several years with good success is STA-BIL® Fuel Stabilizer. STA-BIL® is a blend of scientific additives, all of which act together to prevent fuel from undergoing degradation and oxidation during storage or while being used infrequently. This product acts as a protective wrapper around fuel molecules so they cannot combine with oxygen or other molecules to form new "bad actor" molecules. The "sweetened" fuel can be stored for long periods and perform its job thereafter as though it had just been freshly pumped into the fuel tank straight from the refinery pipeline. The fuel must be treated each year for long-life storage.

Gasoline vs. Diesel Fuel

Present day gasoline has an average storage life of three to four months in above ground tanks if untreated. Underground storage tanks might add an additional two months life. Treated gasoline stored underground extends its storage life twelve to fourteen months, and then another year if it is treated again.

Diesel fuel stored underground will have an untreated storage life of eight months to one year. Diesel fuel treated with STA-BIL® Diesel Treatment will extend the storage life to two years when stored in

Figure 11-24 *The AIR 303 wind generator made by Southwest Windpower Company and is available from Alternative Energy Engineering in Redway, CA. The unit is especially designed for operation in areas of medium to low velocity winds (9-15 mph). AIR 303 produces 12 volts at up to 9 amps/lOO watts in a 15-mph wind.*

underground tanks. Diesel fuel is subject to the ravages of both fuel oxidation and microorganism growth (i.e., algae). Diesel STA-BIL® contains both an effective fuel stabilizer and a powerful EPA registered microbicide to stop algae growth. (See Figure 11-23)

Wind and Water Powered Generators

If we take an open-minded look at power generation and consider what means are required to produce desired results, some interesting alternatives come into view. The purpose of a liquid fueled motor, for example, is simply to produce rotary motion for a generator to convert into electrical energy. Assuming the generator's pulley could be turned by wind or water power, a comparable amount of electrical energy would be produced without the expenditure for fuel.

The obvious requirements here are being in a constantly windy area or near a flowing steam or waterfall. In the past, creative individuals have investigated both of these techniques using homemade items; and many achieved surprisingly good results. Their concept generally equated to using a propeller blade or paddle wheel to turn a drive shaft and large pulley and transferring the motion via a drive belt to a generator with a small pulley. Inasmuch as wind and water flow was not always available, batteries were used for storing the energy for later use. Today, several companies manufacture wind and water powered generating systems. They range in size from 300 watts to 3,000 watts and more, plus several generators can be wired in parallel to provide very large amounts of power. Wind or water powered units are a viable method for obtaining electrical energy to operate various types of communications

Figure 11-25 *Whisper Wind Generator produced by Whisper Company. Unit will start producing power in 7-mph wind (50 watts), and produce 1000 watts of energy in a 27-mph wind.*

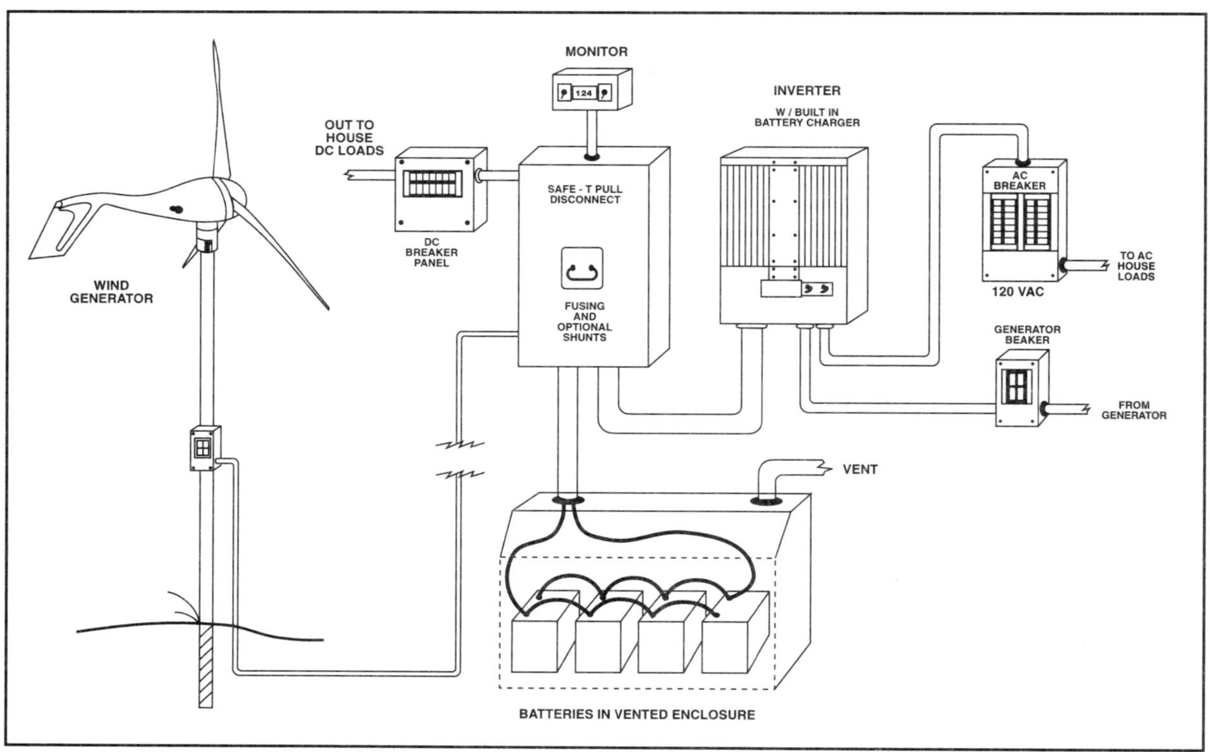

Figure 11-26 *A sample wind generator system shown with component units and hookup. This system produces 120 VAC and 12 volts DC for DC loads. Shown is a possible generator input to maintain battery banks in no-wind conditions.*

equipment and even entire dwellings. They are also superb "backup units" for other alternate energy systems such as solar power.

If you live in a coastal area, on a mountain, or on an open plain where average wind speed is 12 mph or greater, installing a wind generator system holds merit. Commercially made generators typically use two- or three-blade propellers and a modified alternator to produce 12-volts DC. The related current (and output wattage) is, understandably, influenced by wind velocity. A storage battery or bank of storage batteries are charged by the wind generator, and an optional controller (also commercially available) is used to prevent overcharging. Through intelligent planning or working with a commercial supplier, one can usually find the perfect item(s) for a particular site and

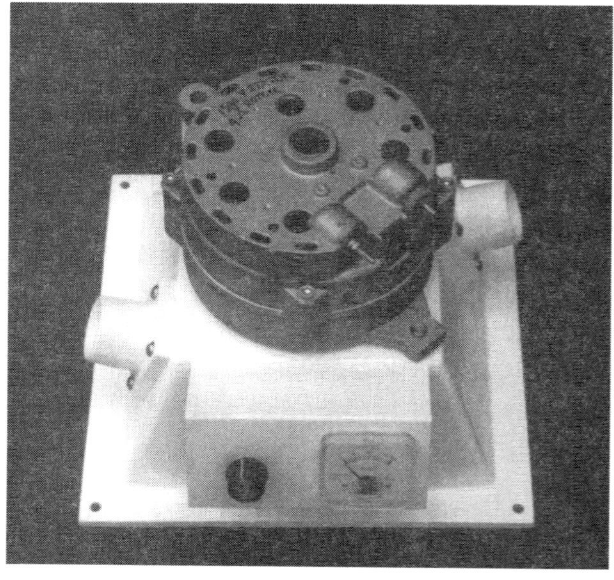

Figure 11-27 *The Harris hydroelectric generator by the Harris Company develops 12 volts at 30 amps/360 watts when mated with a Pelton wheel turbine and powered with water diverted from a small spring or creek.*

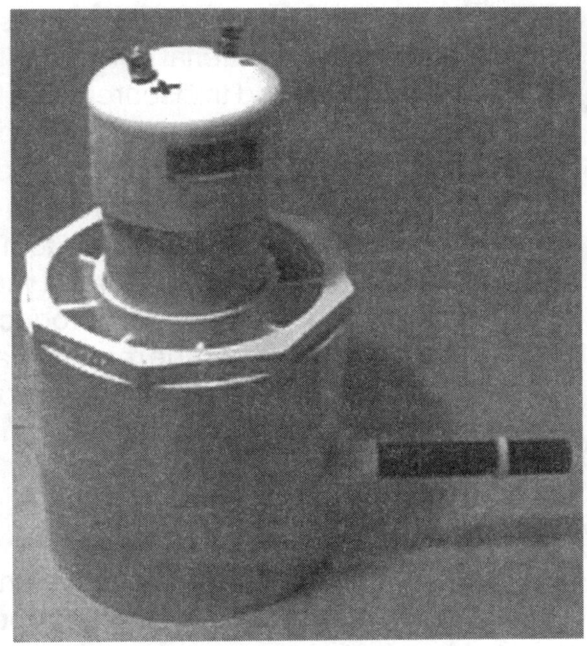

Figure 11-28 *"Lil Otto" mini-generator produces more than enough energy for charging batteries from a very small flow of water. The unit is ideal for powering communications equipment or charging their batteries.*

location. Two impressive wind generators are the "AIR" wind module shown in Figure 11-24 and the Whisper Wind generator shown in Figure 11-25. Both of these generators mount on a single upright pole and include self-protection or tilt capability for surviving 100-mph plus winds. A typical wind system is shown in Figure 11-26.

Assuming water flow in a stream or waterfall can be safely and intelligently accessed and diverted to a head-in point, the use of a small hydroelectric generator holds merit. Caution must be exercised, however, to avoid upsetting nature's balance for wildlife or producing a possible downstream flood potential if problems arise. Using a bypass gate, along with professional guidance, is recommended for water power systems. Fortunately, dealers of alternate energy products are congenial and capable of providing such guidance.

Typically, commercially available hydro generators produce between 50 and 1500 watts of power. Their concept of operation involves concentrating part of a stream's water flow into a specially designed turbine which rotates a modified automobile generator or alternator. The generator or alternator then outputs 12 volts DC, which is used to charge storage batteries. Hydro power may not be feasible for daily use in many locations; however, it may prove quite beneficial for "backing up" other natural energy sources such as solar power.

One of the more impressive high power hydro generators is the Harris Pelton unit shown in Figure 11-27. This generator is available in 350 to 1,500 watt versions.

For small applications requiring 12 to 14 volts DC at 3 to 4 amperes (like charging a small battery to power a 25- or 50-watt radio),

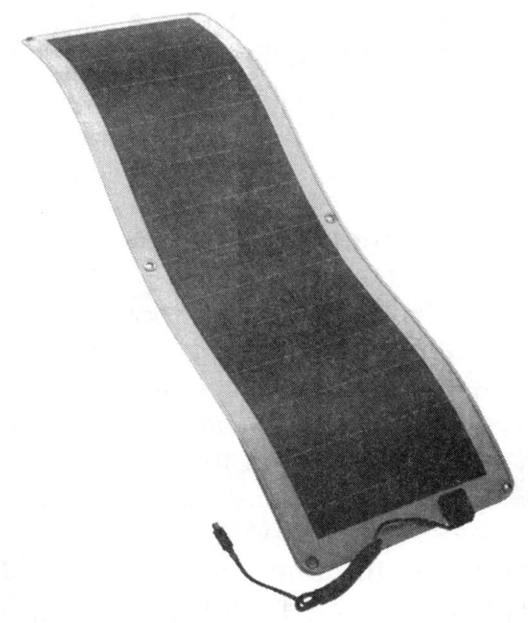

Figure 11-29 *A flexible solar module available from Uni-Solar Systems Corporation produces 5.5 watts of power from sun energy, may be mounted on curved surfaces, and is perfect for charging the batteries of small communications systems.*

Figure 11-30 By combining solar panels in a series and parallel arrangement, a large amount of energy is available. Solar system shown here will last for over 20 years, and requires little maintenance.

Harris Pelton sells a compact "Lil Otto" (Figure 11-28). This mini-turbine and generator is perfect for use with basic survival communications systems, and indicates lower cost mini-units are always available for specific applications. Investigating monthly magazines to determine "today's best buys" is the key to finding them.

Solar Energy Panels

Surely the most well-known form of alternate energy is solar power, or photovoltaic energy. Indeed, the concept of producing electrical energy from the sun's rays dates back several decades and has proven worthy for powering both communications equipment and small to medium size dwellings. Solar panels do not produce the high current obtainable from commercial generators; however, using them to charge one or more storage batteries and striving for energy efficiency is vital.

Simply stated, solar cells use a silicon type material covered with a protective and transparent shield to convert photon energy from the sun into electricity. Groups of small cells are wired in series to obtain a desired voltage (usually 12 to 14 volts DC), then wired in parallel with other groups of cells to produce more current (amperes). Solar panels, comprised of numerous solar cells, are now sold in an endless array of sizes to fit various needs.

Solar panels do not have any moving parts, and do not wear out with use. Manufacturers usually warranty the panels for 10 to 12 years. Factually, their life expectancy is over 20 years. The only maintenance required is periodically cleaning their surface and possibly adjusting their tilt angle twice a year for maximum exposure to the sun. Solar panels are available in all sizes, from small one by two foot, 5-watt versions to four by two foot, 25-watt versions and larger (see Figure 11-29, Figure 11-30 and Figure 11-31).

Figure 11-31 A large array of solar panels is capable of charging a large bank of batteries used to power an entire dwelling.

Figure 11-32 *Two-panel set for emergency communications system. These can be from 55 watts to 75 watts each which will charge two deep-cycle 12-volt DC batteries in the system.*

Solar panels (photovoltaic modules) convert sunlight into electricity and use wire to carry the electricity to batteries to store this energy until it is needed. Before the electricity reaches the batteries, it passes through a controlling regulator called a charge controller. This controller will shut off the flow when the batteries are full.

For some uses, DC (direct current) can be used directly from the batteries to power items such as 12-volt DC lights, radios, low-power transmitters, CB units, ham radio receivers and transmitters, etc. Most common household appliances use alternating current. DC electrical energy can be transformed into AC output by using a device called an inverter and then used as 110-volt AC current .

Most areas of the United States have enough sun exposure to produce ample electrical energy to support average emergency needs. The yearly average equivalent in sun hours per day for the United States is shown in Figure 11-34.

Sunlight intensity is measured in equivalent full sun hours. One hour of maximum or 100% sunshine received by a solar panel equals one equivalent full sun hour (i.e., 14 hours a day). This site may only receive six hours of equivalent full sun. There are two reasons for this. First, is reflection due to the high angle of the sun in relation to your solar panels. The second is also due to the high angle and the amount of the earth's atmosphere that the sunlight must pass through. When the sun is overhead, the light

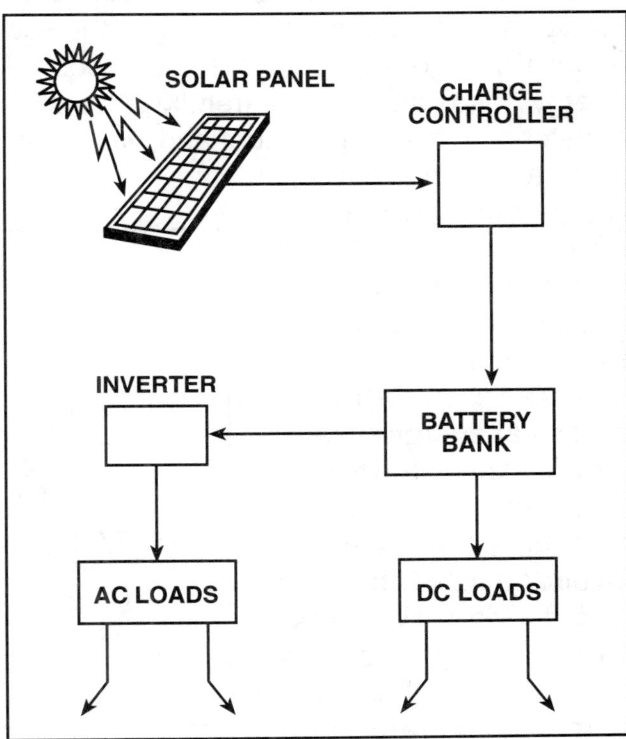

Figure 11-33 *A block diagram of a basic solar energy system producing both DC and AC current for many uses.*

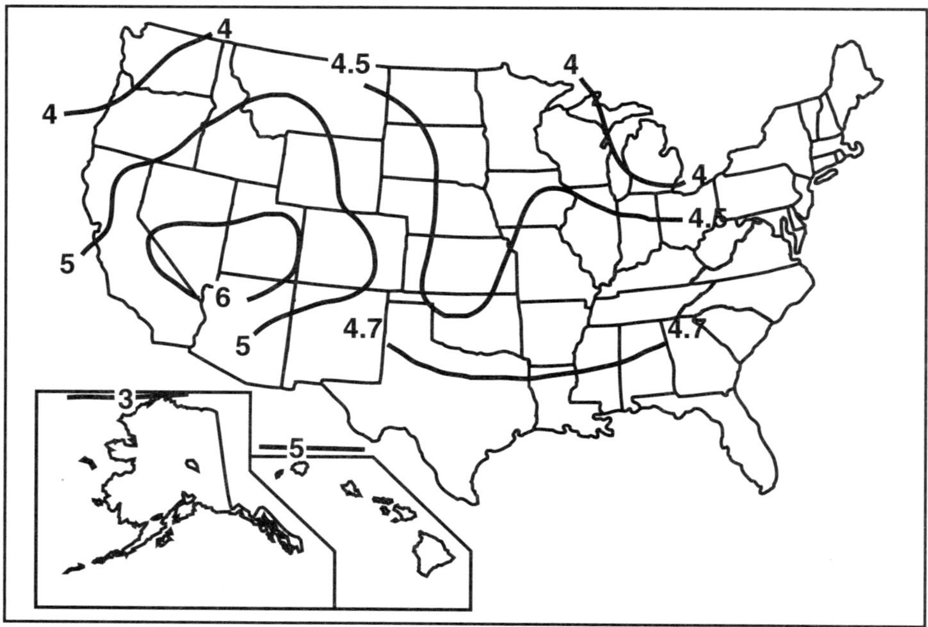

Figure 11-34 *The yearly average equivalent sun hours per day in the United States.*

is passing through the least amount of atmosphere. Early or late in the day, the sunlight is passing through much more of the atmosphere which cuts down the effective amount of energy to the solar panels. (See Figure 11-35 for the seasonal path of the sun which affects solar output.) Your total electrical energy needs should be carefully calculated to determine the total number and wattage of your total solar array. Also, your total battery storage capacity should match your solar array output as closely as possible. Figure 11-36 shows an example 1,000-watt solar energy system using the following components: solar panels, power center, inverter unit, AC breaker panel, battery bank, wiring hookup, and a vented battery box. This sample system can supply some selective DC power loads as needed.

This example system's power production will yield 1,800 watt hours daily from an 8-hour equivalent sun day and the battery storage is 6 volts with 300 to 370 ampere hours. The inverter unit could range from 1,500 to 3,600 watt modified sine wave with built-in battery charger and transfer switch for standby auxiliary power. The system's units can be housed in a dwelling's basement, utility area, or power shed for weather protection.

So, what is the overall best emergency or alternate energy source or system? That depends solely upon your personal needs and budget. Simple alternate energy systems like a single solar panel or portable generator charging a 10 amp hour gel cell or 160 amp-hour lead-acid battery are inexpensive and

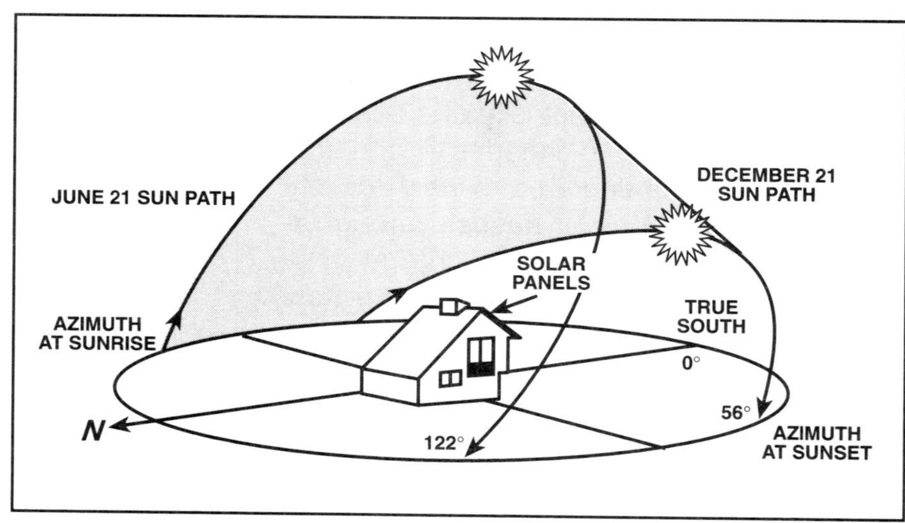

Figure 11-35 *This diagram illustrates the path of the sun over varying seasons. Remember when selecting a site for your solar modules to pick a spot that is clear of shade from a minimum of 10 a.m. to 2 p.m. on December 21st.*

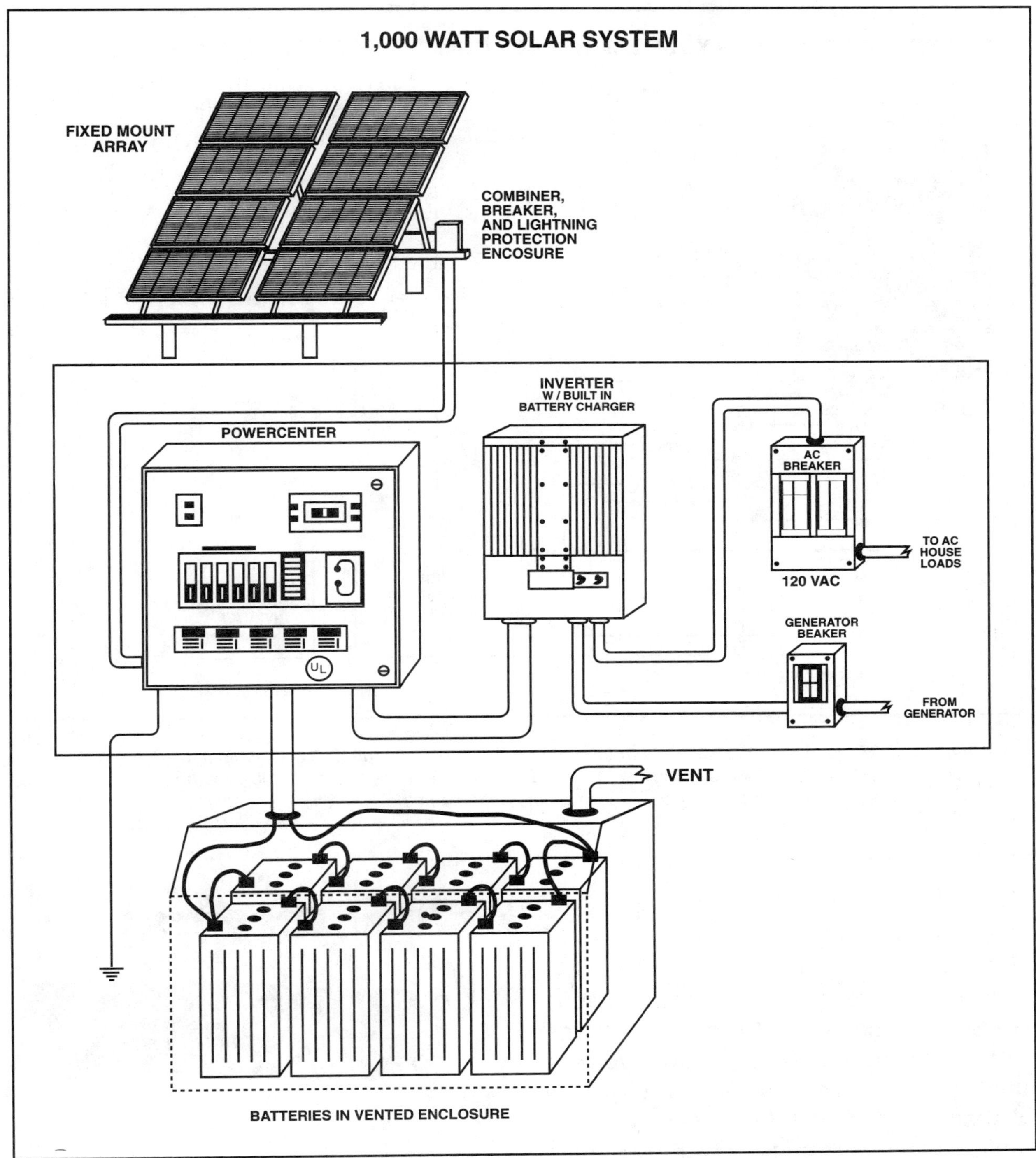

Figure 11-36 *A 1,000 watt solar system—24 volts for primary 110 volt AC loads with selectable DC loads as needed.*

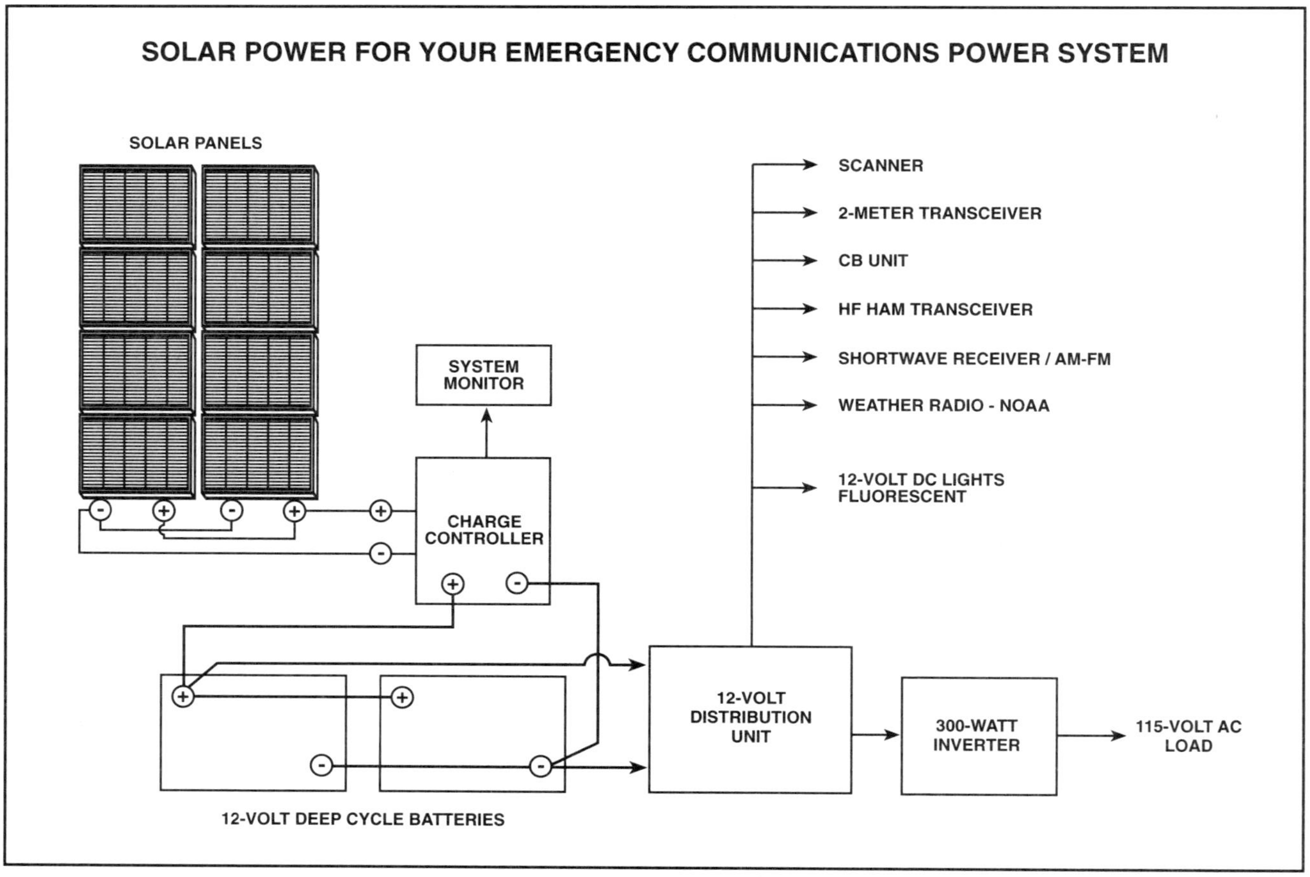

SOLAR POWER FOR YOUR EMERGENCY COMMUNICATIONS POWER SYSTEM

SOLAR PANELS

SYSTEM MONITOR

CHARGE CONTROLLER

SCANNER

2-METER TRANSCEIVER

CB UNIT

HF HAM TRANSCEIVER

SHORTWAVE RECEIVER / AM-FM

WEATHER RADIO - NOAA

12-VOLT DC LIGHTS FLUORESCENT

12-VOLT DISTRIBUTION UNIT

300-WATT INVERTER

115-VOLT AC LOAD

12-VOLT DEEP CYCLE BATTERIES

Figure 11-37 *A simple solar system to support all needed communications equipment. The power output will be 400 amp-hours using two 12-volt DC deep cycle batteries of 200 amp-hours and two 45-watt solar panels. The 115-volt AC inverter is not needed, but the 115-volt AC source is nice to have in an emergency.*

attractive for powering most communications gear. Such setups do not require a voltage inverter, and use a diode for low voltage shutoff resulting in maximum energy efficiency and cost-effectiveness. Many people prefer building a large system to power their entire dwelling; however, this chapter presented a full overview of options. It should thus prove beneficial for addressing both present and future preparedness needs.

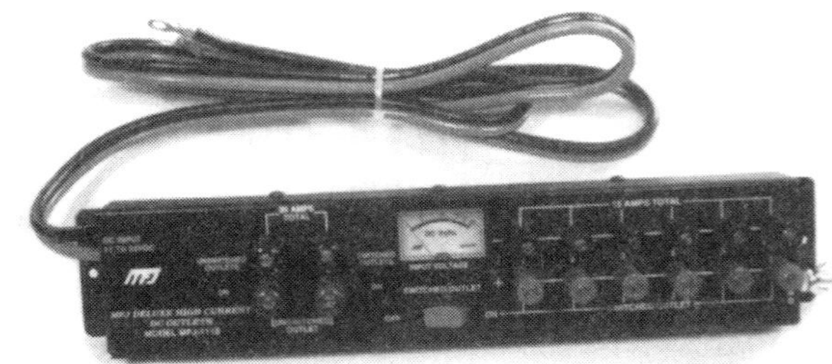

Figure 11-38 *The MFJ 12-volt metering and distribution unit can be used in the solar system shown in Figure 11-34. This emergency communications power system will power a scanner, 2-meter transceiver, CB unit, HF ham transceiver, shortwave receiver, NOAA weather radio, and fluorescent lights.*

CHAPTER TWELVE

Satellite: Your Link to Alternative TV and Audio News and Programming

Backyard satellite systems can provide an unlimited amount of alternative news from many sources and all parts of the world. We are outlining the larger dish systems you have seen for many years. The dish (antenna) measures 7½ to 10 feet in diameter and is referred to as a C-Band system.

The newer small systems known as Direct TV™, Primestar™, USSB™, and others using 18- to 24-inch antennas will not carry the type of services we want to receive. These small pizza-pan size systems are fixed on one satellite. This satellite carries only the selected programming for the services or channels that they sell on a yearly subscription basis. These small dish systems will not receive the satellite channels we need for alternative news and programs.

The larger dish systems (C-Band) can automatically move to any one of the 32 satellites (Figure 12-1). Each of these C-Band satellites has at least 24 channels for a total channel selection of over 768 possible channels—of these over 50% are free and are available without subscription fees or charges of any kind.

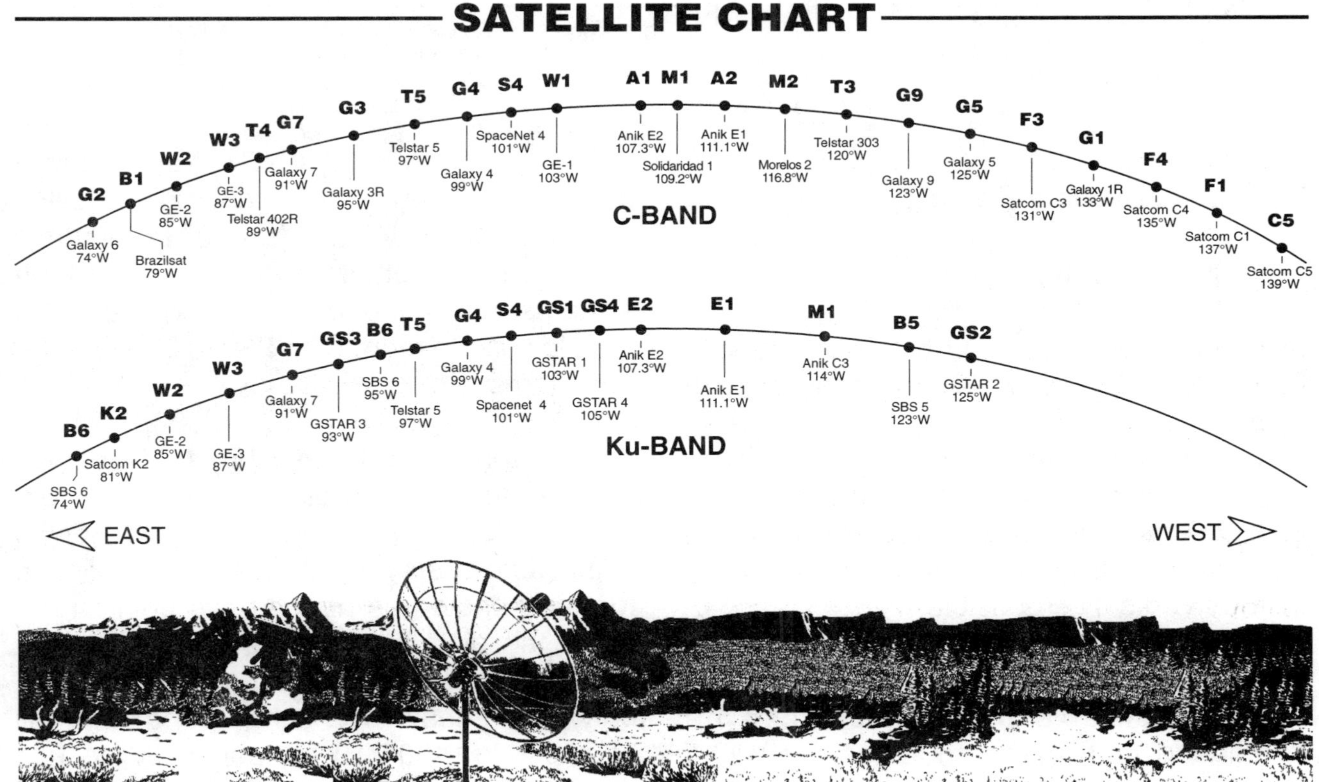

Figure 12-1 *At publishing time, satellite chart showing locations of domestic C-Band and Ku-Band satellites in Region 2.*

How to Set Up an Inexpensive Satellite System to Receive Alternative Programming

Sources of Used Satellite Equipment I have set up several small dish (6 to 7½ foot) audio-only systems in the past for several hundred dollars each by using good-quality used satellite equipment purchased from local satellite dealers. Several dealers gave me two dishes (antennas) just for hauling them away. Another dish and satellite receiver was obtained at no cost from a person who was no longer using them. The dish only had to be removed from his backyard!

Canvas your local satellite dealers, but do not call them. Visit dealers at their locations at a time when they are not busy; mornings seem to be the best time.

Briefly explain to them that you are looking for some used satellite equipment to receive the following audio only: C-Band audio.

Figure 12-2 A used, 8-foot solid fiberglass satellite dish (antenna) built in sections.

Needed Equipment

Dish or Antenna A dish (antenna) of 6 to 7½ feet will give good audio reception. The dish could be larger than 6 feet if available free or for little cost. A 10-foot mesh or solid dish would be great.

The dish should be mounted on a 3½-inch O.D. steel pole set in concrete below the frost line. Use a carpenter's level to make the pole as vertical as possible.

A new, late-manufacture dish (6 feet in diameter) with a polar mount can be purchased from many sources for as little as $199.

Feedhorn and LNB Again, most dealers have many used feedhorns and LNB's that have been taken off satellite systems when their customers have upgraded their systems. A working feedhorn and used LNB should cost no more than $30 to $35 tops for both items.

Coax Cable from Dish to Receiver This is the only part of the satellite system that should not be old or used. The RG-6U coax cable costs only about 28 to 35 cents a foot and should be new. At this point, try to place the dish as close as possible to the satellite receiver's location. You should use only a high-quality RG-6U type coax that will carry 950 to 1450 MHz frequencies. Also, use proper "F" connectors on each end of this coax feedline. Do not use any surplus or cable TV coax cables—these will not work.

Satellite Receiver The satellite receiver (non-IRD) is the control center of this system. It allows you to select satellites, transponders (channels), and to tune the audio subcarriers. The receiver also furnishes the voltage to power the LNB at the dish. This voltage is carried up the coax cable.

Remember, your audio-only system does not require the use of an IRD-type receiver or a decoder module as the audio channel you will be using is not scrambled or encoded in any way—it is in the clear and free!

Since 1982, hundreds of thousands of non-IRD satellite receivers were manufactured and are still around—you just have to hunt for them—they are out there. The average price should be from $25 to $45 tops.

Figure 12-3 *A 7½ foot sectional mesh dish (antenna). This lightweight dish can be disassembled and transported.*

You need a non-IRD receiver that accepts a 950-1450 MHz input from the dish/LNB. Good sources for used satellite equipment are often found in the ads in the small weekly newspapers and shopping papers, or you may wish to place an ad of your own such as: *NEEDED, USED SATELLITE EQUIPMENT, etc.*

Please Note: Do not buy any very early 70-MHZ input receivers. Look only for 950-1450 MHz input models.

TIPS:

(a) When buying used equipment, bring cash when dealing with the satellite dealer. Don't hesitate to tell the dealer your purchase will be in cash.

(b) Your audio satellite system will also receive many unscrambled video services on the satellites. To date, I count over 140 free TV channels that are available to dish owners.

(c) When you get your satellite system up and running, look at all of the satellites and transponders. You will find many other audio programs that you will enjoy. See published satellite guides for locations.

(d) The addition of an inexpensive dish actuator to move your dish to other satellites should be considered. They are very reasonable costwise.

(e) Complete, new satellite systems are available from several companies for $500 to $600.

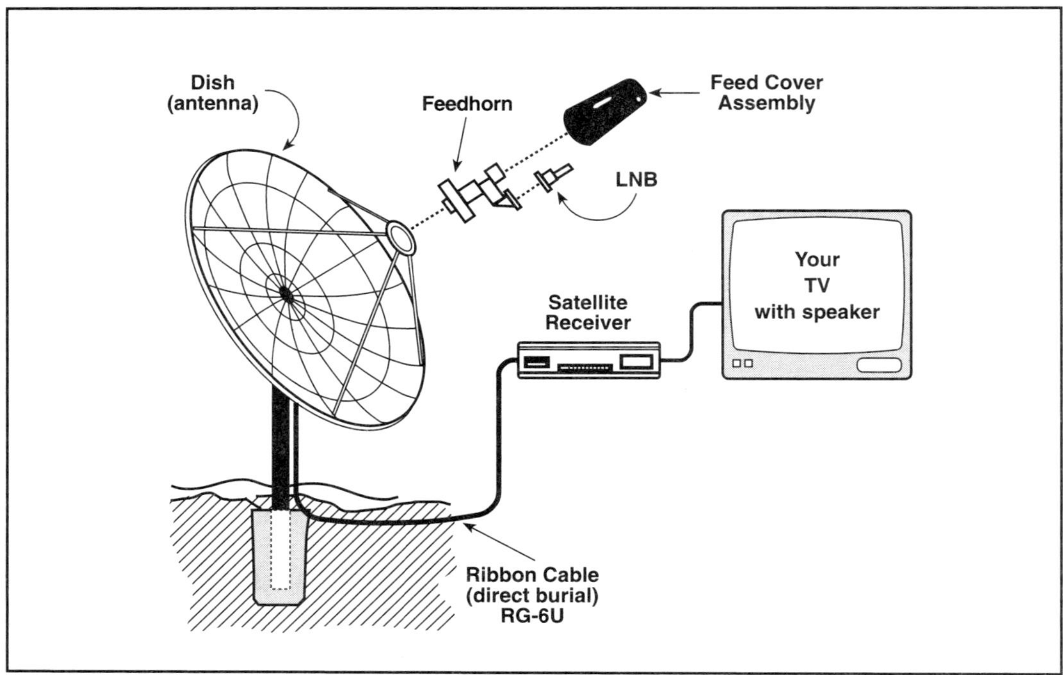

Figure 12-4 *A typical satellite system with components. Shown with TV receiver for audio and TV reception on C-Band.*

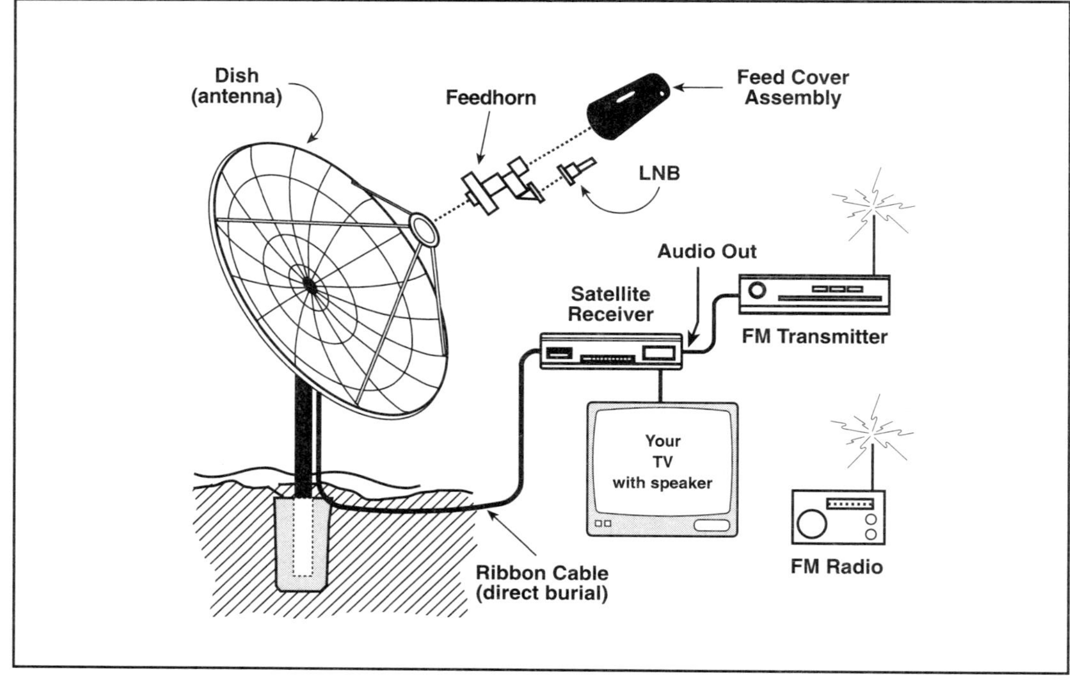

Figure 12-5 *A satellite system with FM transmitter for area coverage of programming broadcast over satellite. This system will broadcast over one-half mile.*

Run Your Own Radio Station

After your audio satellite system is up and running, it is a simple matter to hook up the audio output of the system to a small legal (100 MW) FM transmitter such as the FM-10A that is available in kit form from various electronics dealers. These FM transmitter kits will be offered in built and tested versions if the demand for the wired unit warrants production. The FM transmitter kits are not difficult to build if you know how to do electronic type of soldering. The kit's construction manual is very clear and easy to understand.

The FM transmitter power will cover an entire neighborhood and can be tuned in by any FM radio up to one mile away.

Stereo!

That's right, we said STEREO! Just connect the audio output from your satellite receiver to the FM-10A's line level inputs and you're on-the-air with a school, dorm, camp, home, church, or neighborhood FM station. The FM-10A has plenty of power and the manual goes into great detail outlining all the aspects of antennas, transmitting range, and the FCC rules and regulations. You'll be amazed at the exceptional audio quality of the FM-10A. In fact, your station will sound better than most others on the dial! This is because in their quest for ever higher ratings, FM broadcasters crank up the average level of their modulating signal in order to sound "louder." This heavy compression produces a noticeably more muddy and constrained sound than your signal.

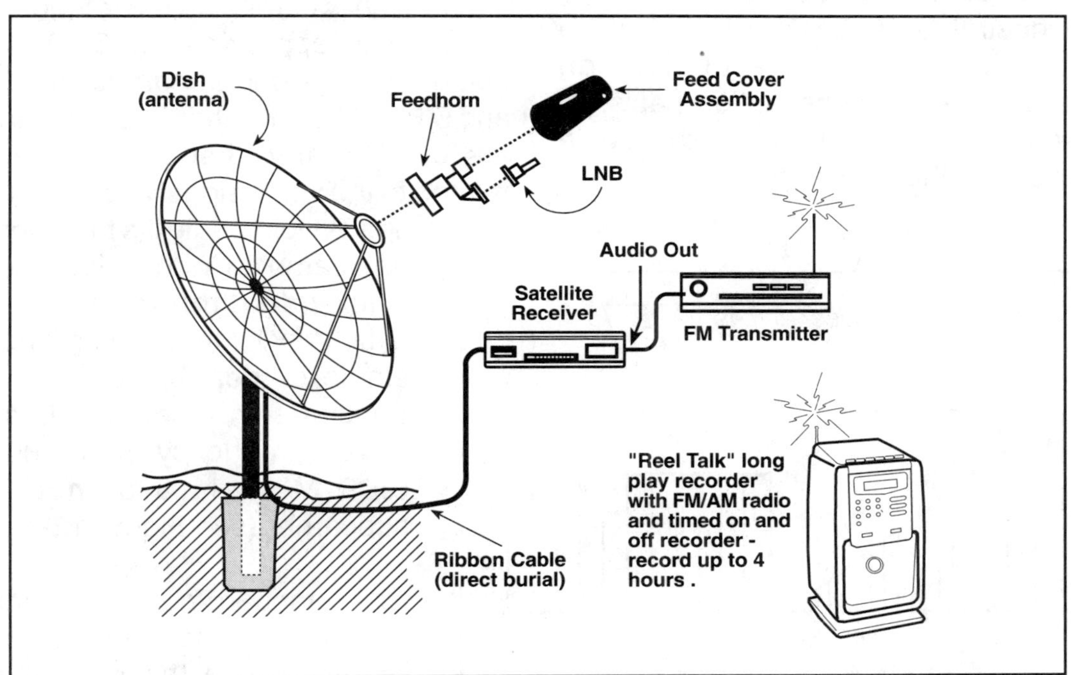

Figure 12-6 *An illustration of the use of an extended play cassette recorder to record programming from the satellites. The use of the "ReelTalk" long-play recorder with timed on and off feature will record up to four hours on one side of a 110 size tape.*

Fully Utilizing Your Satellite System for More Alternative News and Programming

Your home dish can provide countless hours of listening pleasure. In addition to satellite TV, you have access to many audio-only subcarriers that offer music, news, religious programs, and talk shows. Hundreds more radio stations, including those that carry collegiate and professional sports, are available with an inexpensive Single Channel Per Carrier (SCPC) receiver. Satellite dish owners should investigate this fascinating area of cost-free audio subcarriers and SCPC audio services. Most dealers, when selling satellite systems, do not fully explain these added features to their customers.

Following is an outline of all of these available audio services. These services are free and are not scrambled. A complete listing of these desirable services is listed in the "Satellite Radio Guide"®, a publication available on a yearly subscription basis. At the present time, the Satellite Radio Guide lists over 800 of these services.

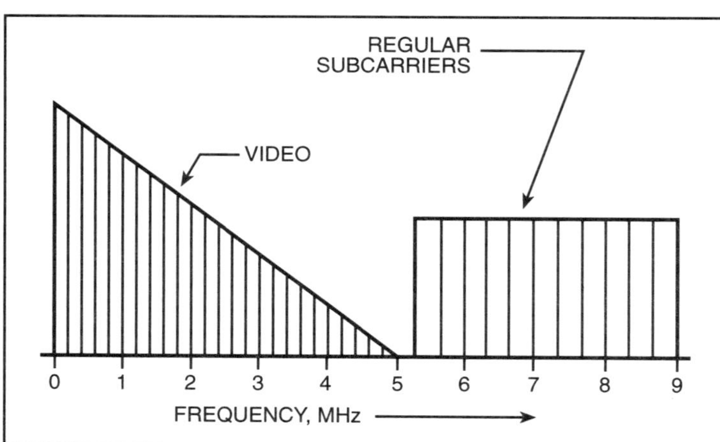

Figure 12-7 *The standard subcarrier frequency plan. 6.8 MHz for video sound with 8 more 15 MHz audio channels.*

Figure 12-8 *Universal SC-50 Subcarrier—FM² Audio Receiver receives all FM² and audio subcarriers—100 kHz to 9 MHz. Full featured audio services, music, all sports, talk shows, news, religious programming, major radio stations, variety, public radio plus many other services, no fees. The SC-50 audio subcarrier receiver will work with all home satellite systems, 3-minute hookup, simple and quick to tune, 16-character display, 50-channel memory bank, direct frequency readout, covers all FM² and audio subcarrier channels, hundreds of free programming channels.*

The Basic Satellite Formats

Standard Audio Subcarriers

The standard audio subcarrier is the audio section of a normal video satellite transponder (unscrambled). The video information is carried on the frequencies from 0.0 MHz to approximately 5.00 MHz. The audio for the video is carried on 6.20 MHz and 6.80 MHz. Any other frequencies can, and are, used to carry other audio subcarriers not related to the video signal in the 5.00- to 9.00-MHz range. A typical satellite transponder with unscrambled video, has one carrier. The video information and the subcarrier information both ride on this single carrier. To sum this up, any transmission mode that is shown to have a frequency from 5.00 MHz to 9.00 MHz, is a standard audio subcarrier (see Figure 12-7).

Standard audio subcarriers can be carried in mono or stereo formats received with the regular satellite receiver and the Universal SC-50 Subcarrier/FM² Audio Receiver (see Figure 12-8).

Another Audio Subcarrier: FM² - FM/FM, Subcarriers Below Video, 0-5 MHz

Now that we have covered the well-known standard audio subcarriers on the satellites (Audio Subcarriers Above Video), a question arises. What if the frequencies below 5 MHz were not used or needed for video transmission? Is it possible to load this unused area with additional audio subcarriers? The answer is yes. In fact, this is the reason we refer to these lower frequency audio subcarriers as subcarriers below the video frequency area, that is, 0-5 MHz (the normal video frequency area).

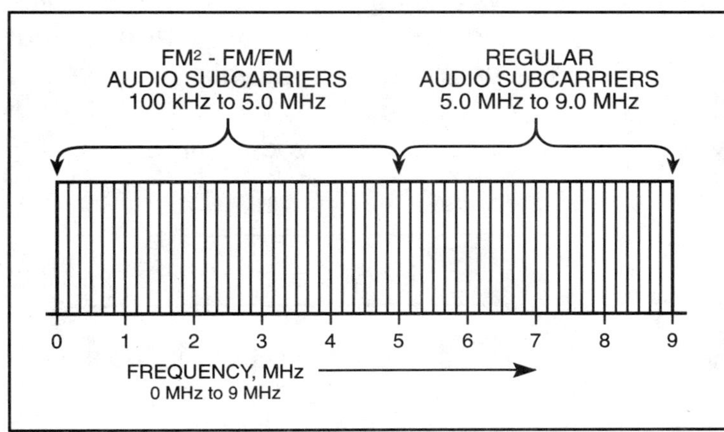

Figure 12-9 *A nonstandard transponder, no video, all subcarriers. Many narrow signals and subcarriers combined. The subcarriers from 0 MHz to 5 MHz are referred to as FM² or FM/FM transponders.*

Several years ago, a company started to use these below-video audio subcarriers by leasing full transponders on several satellites exclusively for the transmission of audio subcarrier service over the transponder's entire frequency range of 0-9 MHz. These transponders do not normally carry any video signals because the frequency space normally used to carry the video signal is used for the transmission of additional audio subcarriers.

Your regular satellite receiver will not receive FM² channels and services. The Universal SC-50 Subcarrier and FM² receiver will receive all FM² channels and subcarrier channels. (See Figure 12-8)

What Is SCPC and What Does SCPC Mean?

Many satellite users confuse SCPC with audio subcarriers. The two are distinctly different in the following way: The family of audio subcarrier transmissions are linked to the satellite video waveform and use this method to carry the subcarrier signal, one primary carrier.

The SCPC signals (channels) have their own independent carriers and each has its own spot frequency throughout the transponder on which they are being carried.

SCPC means Single Channel Per Carrier, or Single Carrier Per Channel. Both seem to be correct, as each narrow band channel has its own carrier which is modulated. An entire transponder is usually used for SCPC transmission.

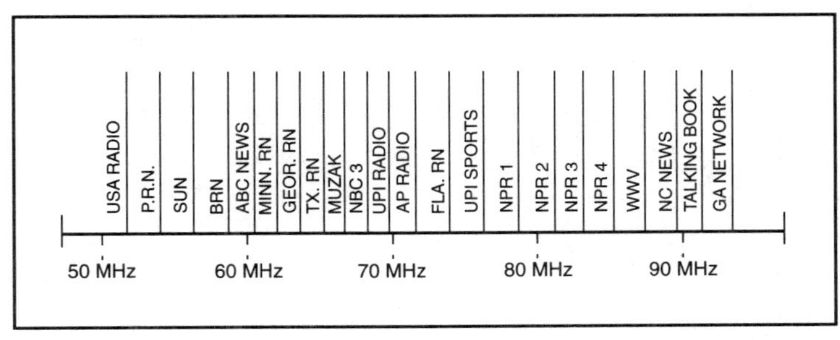

Figure 12-10 *Typical (not actual) SCPC single transponder service assignments*

The entire transponder frequency from 50-90 MHz is divided into many low-level carriers which are individually modulated on set frequencies.

In general, all SCPC signals are classified as narrow bandwidth signals. However, some are very narrow—narrow is 7.5 kHz and wideband is 15 kHz. The regular satellite receiver will not receive SCPC channels. You must have an SCPC receiver which is easy to install in your system. The Universal SCPC-200 will receive all SCPC channels. (See Figure 12-11)

Figure 12-11 Universal SCPC-200 Audio Receiver provides full commercial features, direct frequency readout, easy direct frequency and transponder tuning - 50 to 90 MHz (LCD display), large memory bank - 50 channels, C and Ku-Band agile - 950 - 1450 MHz, automatic LNB drift compensation (ADC), companding, 1:1, 2:1, 3:1 (automatic), wide and narrow bandwidth, automatic tuning indicators, digital frequency lock-on (DFL), service name on LCD display, microprocessor frequency display, speaker and line outputs, high quality audio, commercial digital synthesizer, 6-button keypad for fast tuning, baseband 70 MHz output, built in U.S.A. by the leading SCPC manufacturer, full 16 character LCD display, and does not disable video when in use.

"Satellite Radio Guide"™ Lists All Satellite Radio Channels

This guide is the only complete and accurate quarterly listing of all audio services on all satellites. The "Satellite Radio Guide"™ includes satellite audio subcarrier services above video (5.0 MHz to 9 MHz), satellite audio subcarrier services below video (100 kHz to 5.0 MHz - FM^2 - FM/FM), a complete SCPC channel listing (all satellites, Region 2), new entries, and channel changes. This is the most complete listing in the industry. It also covers other interesting satellite services, talk shows, alternative news, weather facsimile photos, press facsimile services, NOAA weather photos, and late-breaking world news and news services. The "Satellite Radio Guide"™ is published quarterly (four times a year). It is available on a yearly subscription basis. Each edition is sent by first class mail so it is received on time.

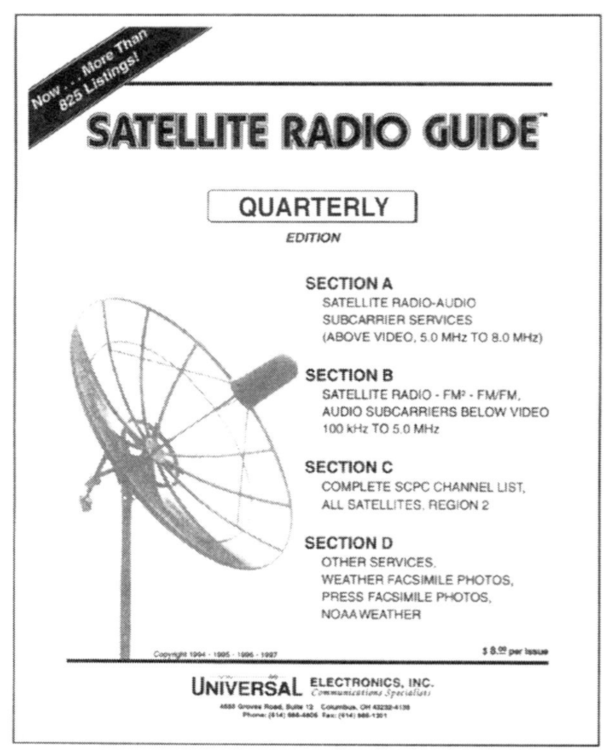

Figure 12-12 The Satellite Radio Guide published by Universal Electronics, Inc. Phone: 614 - 866-4605

Other Publications on Satellite Reception

Tune to Satellite Radio on Your Satellite System

"Satellite Radio"® is the first complete, up-to-date book on the audio services carried on most domestic C- and Ku-Band satellites. These audio services can be received by most home satellite systems (TVRO), sometimes without additional equipment. Any additional equipment is usually low in cost. The audio programming is high quality in transmission and in content—music, talk shows, radio programming, radio networks, all news services, business services, weather services, all major sports are available on "Satellite Radio." Hundreds of services are "up there" and "Satellite Radio" shows you how to receive them.

"Satellite Radio" also has an enclosed satellite audio services guide which is not bound in the book and is updated quarterly. This is the most complete satellite audio guide in the industry. A one-year subscription is available with the purchase of this book.

The Satellite Radio Guide Section of "Satellite Radio" is updated because there are regular major changes in programming. The Satellite Radio Guide feature gives you all the needed information to enjoy "Satellite Radio." 103 Pages.

Install, Aim and Repair Your Satellite TV System

A shortened version of the book Home Satellite TV Installing and Troubleshooting Manual. This booklet explores how to install and troubleshoot a home satellite TV system. This clearly written book shows any layman how to install a satellite TV system, aim the dish at the satellites, and repair the system if problems should arise. Written for consumers. 64 Pages.

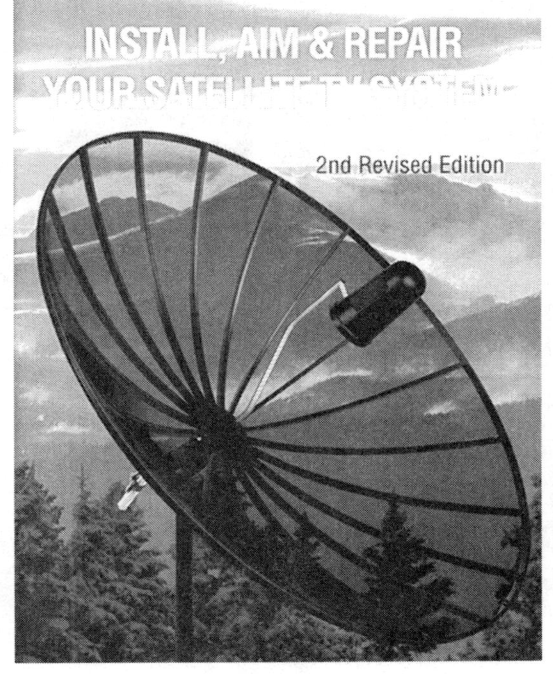

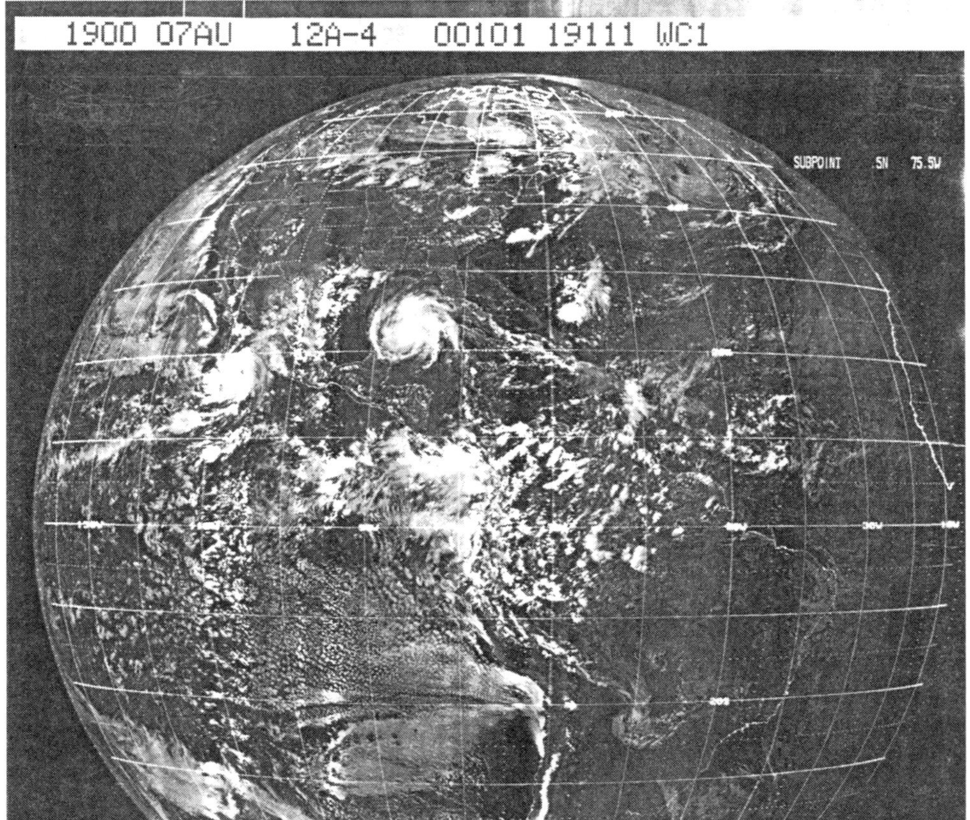

Figure 12-13 *A satellite printout from a weather satellite in space. Note the two major hurricanes, one in the Gulf of Mexico and one on the west coast of Mexico. Weather satellites are one of the most important tools of modern weather forecasting, they afford days of warning before landfall allowing preparation and evacuation of dangerous areas.*

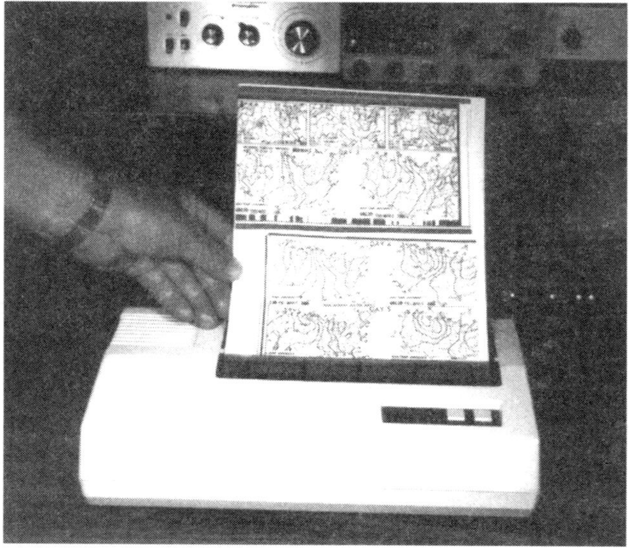

Figure 12-14 *Section weather charts being printed on a standard computer printer. These weather charts are sent via satellite transmissions by the National Weather Services (NWS) and the U.S. Navy. Shortwave radio also carries the same type of weather information. The equipment to receive weather information using you home computer is readily available at a moderate cost.*